T0325088

Secrets of Machine Learning

How It Works and What It Means for You

Secrets of Machine Learning
How It Works and
What It Means for You

Tom Kohn

 **World Scientific**

W JERSEY · LONDON · SINGAPORE · BEIJING · SHANGHAI · HONG KONG · TAIPEI · CHENNAI · TOKYO

Published by

World Scientific Publishing Europe Ltd.

57 Shelton Street, Covent Garden, London WC2H 9HE

Head office: 5 Toh Tuck Link, Singapore 596224

USA office: 27 Warren Street, Suite 401-402, Hackensack, NJ 07601

Library of Congress Cataloging-in-Publication Data

Names: Kohn, Tom, author.

Title: Secrets of machine learning : how it works and what it means for you / Tom Kohn.

Description: [Hackensack] New Jersey : World Scientific, [2024] |
    Includes bibliographical references and index.

Identifiers: LCCN 2023039960 | ISBN 9781800614888 (hardcover) |
    ISBN 9781800615021 (paperback) | ISBN 9781800614895 (ebook for institutions) |
    ISBN 9781800614901 (ebook for individuals)

Subjects: LCSH: Machine learning. | Industry 4.0. | Artificial intelligence--
    Social aspects. | Automation.

Classification: LCC Q325.5 .K64 2024 | DDC 006.3/1--dc23/eng/20240228

LC record available at https://lccn.loc.gov/2023039960

British Library Cataloguing-in-Publication Data

A catalogue record for this book is available from the British Library.

For any available supplementary material, please visit
https://www.worldscientific.com/worldscibooks/10.1142/Q0440#t=suppl

Desk Editors: Logeshwaran Arumugam/Rosie Williamson/Shi Ying Koe

Typeset by Stallion Press
Email: enquiries@stallionpress.com

Printed in Singapore

To Ellen, Leo, and Scout

About the Author

Tom Kohn is a news product manager in Bloomberg LP's News Innovation Lab. Before that, he was a journalist for two decades — in London, Hong Kong, Frankfurt, and New York City. His focus is on the intersection of news, finance, and technology and he has a fascination with machine learning and its real-world applications. His specialisms are automation, financial markets, and data.

Contents

Introduction: It's All Just Pixels

Imagine the world before light bulbs. Now consider how the world changed once they were invented. We're now at a similar time in history. There's a technological watershed taking place that's akin to the Industrial Revolution. And you need to know how and why things are transforming. Machine learning is changing the workplace, it's changing the economy, and it's changing society. Computers can recognize potential lung cancer better than doctors, detect fraud better than bankers, and create fake video that are almost impossible to tell apart from the real thing. Next, machines are likely to write up instant legal documents and transport goods from coast to coast without drivers. But this is a new industry awash with hype. It's crucial to understand what is real and what is fantasy, which predictions about the future are possible, and which are just hot air or wishful thinking. To stay cutting edge in your chosen industry or career, it will help to understand the forces behind these tectonic shifts, as well as some of the details that make this technology work. That's the crux of this book. Two facts are the key to this transition. The amount of data in the world is surging year by year. And computers are getting more and more powerful. This combination of a surge in processing power and the sheer quantities of data available that can be used to train machine learning models is leading to performances that are now possible which were unthought-of 10 years ago. As stated by AI pioneer Andrew Ng:

Artificial intelligence is poised to transform every industry, just as electricity did 100 years ago.[1]

What this means is that anything that can analyze and make sense of all that data is going to be supremely useful. And anyone who can do it will be in demand. That's where data scientists come in. The field is exploding and is predicted to grow further in coming years as more companies seek to become data-driven.[2]

Self-Teaching

One more reason for the proliferation of the different types of learning methods in recent years is that machine learning can teach itself and work without explicit instructions from computer programmers. That's an exciting prospect, not least for programmers themselves. This technology affects everyone. Algorithms determine your credit score, the kind of television you watch, how you shop, and, soon, how your cars will drive themselves. Yet artificial intelligence is still imperfect and riddled with irritations for the user. Computers can't see, read, or understand in any meaningful way, at least not in the way people comprehend it. Self-driving cars were predicted to have been up and running on roads many years ago, and fully autonomous vehicles are still mostly out of reach.

The chapters that follow will explore the origins of machine learning and why it has suddenly proliferated, and cut through the hyperbole surrounding artificial intelligence systems. It is key to understand how this seismic shift fits into the timeline of historical technological breakthroughs. Is this technology really analogous to the invention of the steam engine, the lightbulb, or the transistor? The answer, as will become plain, is yes. But more important for the reader is to describe in practical terms what you need to know about how machine learning works, where it applies, and how companies and workplaces are changing because of it. The following chapters will do that too. Machine learning isn't magic. It can do some amazing things, and the speed at which the technology is changing makes it all the more breathtaking. Also, because we don't know why certain algorithms work, it somehow makes the fact that they produce such accurate results a bit mysterious.

This book explains some of these jargon-laced and complicated terminologies, demystifies the technical aspects, and clears up some of the murkier concepts within machine learning. We describe why this revolution is happening now, and what is so great about it, as well as some of the risks that are emerging. Ethical concerns are only now arising as the technology adapts and progresses. Some academics fret about existential risks as machines gain real intelligence and pose a threat to humanity. Others say this threat is nonsense, and the real risks lie in job losses from automation as systems get smarter and cheaper to adopt. Academics, scientists, and industry practitioners are occasionally in agreement about certain areas of concern, specifically regarding autonomous weapons, also known as killer robots, along with facial recognition and privacy. We examine both in detail.

How We Got Here?

To explain how we got here, we need to go back to the dawn of the industry and examine some key moments along the way in the history of computing in the last 50 years. But it's in the past decade where things really took a quantum leap to allow the current technology to take off. What really caused liftoff, and shifted a lot from theory to practice, was the rise in the availability and amount of data, along with the concurrent increase in processing power. Now these technologies power everything from virtual assistants, to loan approvals and medical diagnostics.

There's no sorcery to machine learning, as we'll see. It really is all just pixels on a screen, binary numbers, or on–off switches. From systems that recommend movies, to search engines, to automatic translation, at heart there's a conceptually simple mechanism that works in zeros and ones. Computers use brute force to crunch through numbers and use statistics and probabilities to infer outputs, which from the outside can lead to answers that can seem amazing. We'll examine how they work in a practical way, helping readers to understand the new world around them to learn the skills needed to use and interact with machine learning systems without fear or confusion. We'll synthesize knowledge from the world's experts in the field and distill the complex concepts into easy-to-digest chapters that cover industries and types of machine learning systems. We'll also cut through some of the bluster and take a skeptical look where needed, helping the reader navigate the complexities of the field. We'll also consider the risks and pitfalls. There are ethical considerations that must be taken into account when building models: There's industry techno-evangelism and unrealistic assumptions about what machine learning is truly capable of. Some academics highlight existential risks. Practitioners say that there's a bigger risk than AIs turning evil: AIs turn competent, but with goals that clash with those of humanity.[3]

Philosopher Nick Bostrom's thought experiment of a paperclip maximizer, where an intelligent machine unwaveringly pursues its assigned task by making the most paperclips it can regardless of human life, is one extreme version of where artificial intelligence might take us.[4] Other scientists in the field dismiss this kind of worry but point to more present ethical concerns such as algorithmic bias perpetuating discrimination, or mass job losses from technology, all of which need discussion. We'll dig into those ethics and their practical applications. Artificial intelligence is a key driver of what World Economic Forum founder Klaus Schwab describes as the Fourth Industrial Revolution,[5] as technologies meld, provoking a profound shift in production, consumption, and transportation. The effect of machine learning will be felt across homes, businesses, schools, and public spaces, promising to solve some of humanity's pressing issues but also throwing up risks, such as algorithm transparency, data privacy, and job displacement.[6] This upheaval also presents opportunities for people and for companies. There are also some more metaphysical questions about whether machines can ever be like us, or think like us. And if they provide responses that make them seem conscious, even though they aren't, does it even matter?

Who Should Read This Book?

Anyone curious about how technology is changing in the workplace and the economy will benefit from learning how machine learning is filtering into different areas. Employees looking to future-proof their own skills, students seeking to get ahead as they enter the job market, and anyone with a general interest in cutting-edge industry. This is not a textbook aimed at computer scientists or technologists, yet they may also find some nuggets of interest here in terms of how different sectors are adopting machine learning and at what speeds. While we will strive to avoid too much technical jargon or mathematics, we will include a list of further reading for anyone interested in diving deeper into the technical side of things, along with a recommended list of books and more specialist learning. This book examines the different industries where machine learning operates and is being adopted, how the different systems work, and how the technologies are changing industry, the economy, and the job market. Many roles are changing in the workplace as companies race to adopt strategies that save them time and money and keep ahead of the competition. New kinds of work and jobs are also springing up as workplaces adapt. We will suggest practical steps to stay abreast of these watershed changes and highlight the skills you need to know in the future. Finally, we'll address some of the ethical concerns with algorithms behind machine learning and the use of big data to power these systems. Giant chatbots suddenly seem to be everywhere in the news, but artificial intelligence is more embedded in our lives than just these. From voice assistants like Apple's Siri and Amazon's Alexa, to movie recommendation systems like those built by Netflix or Amazon Prime, to internet search engines or social media, similar underlying machine-learning technologies are powering our lives. If you want to know how and why, read on.

Chapter 1

Terminology Unjumbled

Terminology in artificial intelligence is confusing. What exactly are machine learning, deep learning, and neural networks, and how are they distinct? In this chapter we unpack why it's crucial to learn what all the definitions mean in order to understand how it all works. Even artificial intelligence itself is a misleading term. We now have a plethora of amazing tools related to speech, text, and images, but there's still a dearth of knowledge about the underlying technologies, which can be boiled down to simple concepts that apply in concrete ways to the real world.

People worry that computers will get too smart and take over the world, but the real problem is that they're too stupid and they've already taken over the world.[1]

— Pedro Domingos

To start with, consider machine learning as a computer system that learns over time and teaches itself. Here's IBM's definition: "Machine learning is a branch of artificial intelligence (AI) and computer science which focuses on the use of data and algorithms to imitate the way that humans learn, gradually improving its accuracy."[2]

So the key difference from earlier types of computer code, where programmers wrote recipes for computers to follow, is that machine learning doesn't just follow instructions, it learns and improves on its own. The Berkeley School of Information describes machine learning as using statistics and optimization to let computers analyze data and find patterns.[3] That's a good way to look at it. Berkeley researchers describe three components of a typical supervised learning model: a decision process, an error function, and an updating (or optimization) process. That means checking on inputs and outputs, seeing what predictions the machine got wrong, and then updating the system to improve next time. The key is that the computer learns without human intervention, so it can uncover insights without instruction from the programmer. But machines aren't really "learning" in any human way.

Artificial Intelligence Is Real Stupidity

There's no such thing as artificial intelligence in the semantic sense. Some scholars argue that there never will be; machines just can't think or reason in the way that humans can. That is not to say that computers can't learn or solve problems, or even handle certain decision-making. Deep learning, neural networks, natural language processing, and generative algorithms are all just self-learning tools that serve specific tasks. While machines can improve at routine chores, they have no wisdom or intentionality. One of the problems with the field is all the definitions; linguistic confusion, as journalism professor Meredith Broussard puts it.[4] The term artificial intelligence, or AI, has one such definitional problem, in that it's misleading as to what the term actually means. The reason this book is focused specifically on machine learning as opposed to artificial intelligence more broadly is that it sets the scope to within clearer boundaries. Artificial intelligence means different things to different people; it's vague and unhelpful. Some people use it as a synonym for machine learning, some as a subset. There's no sentient being in the computer, at least not yet. Artificial intelligence is also commonly understood to mean what computer scientists call Artificial General Intelligence, or AGI. That is distinct from Narrow AI, also known as Weak AI, which is designed to solve a single problem, such as to play a TV game show like *Jeopardy* or to recommend a movie. AGI is a more wide-ranging theoretical intelligence that can act upon multiple types of tasks or problems and even equal human intelligence. Also, it doesn't yet exist. People are getting confused about the term AI.[5]

Apple co-founder Steve Wozniak proposed a hypothetical test for machines to see if they have human intelligence. The test is to go to his house and make a cup of coffee. The idea is simple for humans, who understand the concept of kitchens, know what coffee machines look like, could guess which cupboard the coffee is stored in, and would know to plug in the machine and switch it on. This task is mind-numbingly complex for a machine with no prior knowledge or instructions, or even any concept of what coffee is.[6] Also, no one would want to build a single-purpose coffee-making visitor robot. The way machine intelligence currently works, if you built such a device that succeeded in making coffee, it would unlikely be able to even make tea, let alone do anything else useful.

Learning Machines

Machine learning is more tangible as a concept, although some argue that even learning is a misnomer in this context, as these systems don't learn in the same way as humans. Regardless, there are three broad types of machine learning: supervised learning, unsupervised learning, and reinforcement learning. In supervised learning, models are fed training data with the correct outputs labeled by people. In other words, in a horse classifier model, pictures of horses are tagged "horse" and those without are labeled "not horse." The machine is then able to look at pictures with no labels and identify horses based on this training. More broadly, in supervised learning, the machine examines the training set and goes on to make predictions on its own, based on the annotated data. The "supervision" is in the annotations (not by a human supervisor). Unsupervised learning is where models are given no such annotated guidance. This means the machine has to make its own inferences and figure out the correct outputs on its own. Humans still build the algorithms, feed data into the model, and watch over the results, but they don't pre-annotate the training data. Clearly, it's much harder for machines to deduce the right outputs, or make predictions, if they haven't been given instruction, but the advantage of unsupervised learning is that it can work at a far greater scale, with no need for human experts to label thousands of documents beforehand. The system is "unsupervised," only in that there is no annotated data from which to learn. Reinforcement learning takes it a step further and offers the machine a reward for a correct output, speeding up the learning process. In reinforcement learning, the right classifications or predictions receive positive feedback and the wrong ones get negative feedback. The system adjusts its behavior to try to maximize rewards, hopefully yielding the correct predictions and outputs faster. Reinforcement learning typically works best in closed or bounded systems, like games.

Dumb Machines

Artificial intelligence systems, such as vision or speech recognition software, are tools designed for a specific purpose. They are not artificial in any real sense in that they were designed to exist in the real world to solve real problems. If there's some sense in which they are "artificially" replacing another manual process of doing that thing, then perhaps that can be true, but that's not what is meant by the semantics of AI. Intelligence meanwhile is an ephemeral concept, even for humans. Reasoning, perception, natural language communication, motion and manipulation, learning, representing knowledge, planning, social awareness and skills, and general intelligence are all forms of intelligence.[7] The list helps us understand AI because we can use it to decide if a given system is demonstrating intelligence. And they aren't just now. So intelligence just can't be applied to machines yet. A thing may look intelligent because it repeats and learns and improves, or makes (to us) startling inferences. But machine learning, deep learning, and the rest are really just statistics and probability. There's no understanding in these systems nor is there supposed to be. They are narrow tools for narrow tasks, the opposite of intelligence as we understand it, hence computer scientists' description of, and interest in, AGI. So artificial intelligence is no such thing: AI is neither artificial nor intelligent. Why then was it given the term? Stanford professor John McCarthy coined the phrase artificial intelligence around 1955 at a conference at Dartmouth College. He described it as "the science and engineering of making intelligent machines, especially intelligent computer programs."[8] The way this developed into an academic discipline, and then into a whole industry, came about as computers became more and more entrenched into our lives. But this was followed by a series of AI Winters, with droughts in funding and a concurrent falloff in research. That's where the technologists over-promised what was possible and caused setbacks in research and financial backing to dry up. Part of the problem of the general public's misunderstanding of how AI does what it does is due to the terms involved. That's at risk of happening again.

Clever Humans

Geoffrey Hinton, one of the founding fathers of the industry, has advised against using the term artificial intelligence altogether to avoid confusion with deep learning and machine learning.[9] Hinton specializes in deep learning and was driven by the desire to understand how the brain works. In fact, the term machine learning was coined around the same time as artificial intelligence, by Arthur Samuel, as part of his research around playing checkers in the 1950s.[10] Over the years, it has come to mean something more specific. The industry's rebranding, where it steered artificial intelligence terminology specifically toward machine learning (ostensibly a subset of AI, although that is also disputed), came after a series of crises when the overpromise of the technology gave way to skepticism, which came in waves over several decades. Also, the computing power wasn't ever enough for the theory to become practice. That has now changed. Even now, with the amazing, incredible things that machine learning can do, the applications are all very narrow. Computer vision systems are upset by data they haven't encountered before, like a self-driving car being unable to distinguish a light blue truck from the sky. Text mining is derailed by changes in format or the moving of columns or tables, and speech recognition is hampered by background noise. There's no human-level intelligence in any of these systems, let alone consciousness. Machines can capture how a thing looks or sounds and, over many thousand or million iterations, learn to detect such a thing again in future. But it's really still mostly just a form of complicated pattern matching. There's no understanding or reasoning available yet. This gap is a cause of intense debate over the future of the industry.

So none of these can work as general intelligent systems, which would need to be able to handle levels of uncertainty and ambiguity ways beyond what is built into narrow machine learning models. Why is there so much confusing terminology? Part of it is deliberate to maintain exclusivity, and part of it is because jargon tends to proliferate in specific fields. Also, it's a complex and technical subject. The systems that will be discussed and explained throughout this book — neural networks, deep learning, GANs, RNNs, CNNs, and GPTs — all fulfill specific and narrow requirements. They have specific tasks, are built in a specific way, and are tailored to specific datasets. What we have is a plethora of amazing tools. The term machine learning is preferable to artificial intelligence because it's more precise and bounded. A lack of definitional agreement shouldn't detract from an important truth, that this new technology will affect a lot of aspects of our lives.[11]

What about the incredible achievements and advances in these technologies? Google Translate is no less amazing because it has no inherent intelligence. Apple's Siri or Amazon's Alexa can still be helpful and grow ever more useful as they improve, without knowing or understanding what you mean; Microsoft virtual assistant Cortana too. The same can be said for Amazon or Netflix recommenders, as we'll see in later chapters; the predictions are incredible and growing more relevant. But there's no intelligence behind them or in them. So what is the upshot of these systems for the real world? It's more likely that the near-future state of the art in learning will come in the form of a machine learning virtual assistant that answers questions based on a huge database and which learns and improves over time. This will have far more information at its disposal than any one person but certainly no consciousness or self-awareness. Think of the question-and-answer abilities of the Star Trek Computer rather than the generally intelligent Commander Data or the humanoid robot See-Threepio in *Star Wars*. Such examples of autonomous general intelligence are far off, whereas the supercomputer that gauges your question and answers it intelligently seems pretty close to reality. As far as machine learning systems are concerned, it's more useful, and far more practical, to analyze them topic by topic and industry by industry to see how they each work and how they are changing our lives than theorize about computers gaining consciousness or how soon we will have real general robot helpers. And one of the most prevalent, and successful, uses of machine learning has been in recommendation engines, to which we will turn next.

Chapter 2

Recommendation Nation

Algorithms are behind the recommendation engines that are suddenly popping up everywhere, suggesting what movies you should watch, which products you should buy, and what you should write in emails or texts. In this chapter, we cover what these engines are and why they are now so important to technology companies. There are some advantages to having machines predict and lay out your choices, but there are some disadvantages as well. How will these systems define our future?

When you power up your television and log in to Netflix, Hulu, Amazon Prime, or another streaming service, you're presented with a multitude of suggested movies and TV shows that are curated for you by machines. How do they do this, and why are they so good at it? Titles are broken into categories designed to increase your engagement and to help you find something relevant and interesting to watch. What is the technology driving these recommendations and how does it work? Behind all these movie and TV suggestions are recommender systems, also called recommendation engines. These are giant filters that use machine learning to predict choices that a user would probably like. They work by siphoning a vast array of titles down to those deemed the most relevant, which are then shown to the viewer or listener. This can take the form of music, movies, or products to buy.

In this chapter, we will examine these systems at Netflix for TV and movies, at Amazon for products, movies, and music. We will also look at the types of predictor systems in machine learning and examine how they work.

Robots Are Deciding What You Watch

Machine-learning systems examine historical data and use that to guide future predictions. The more precise the data and the larger the volume, the better the prediction. This explains why these media and technology companies have been improving and expanding their systems at such a breakneck pace in recent years. As more and more users are attracted to their services, the very subscription and engagement with these platforms help generate mountains more data about customers. At the heart of it, recommenders take such data and make inferences that go far beyond human suggestions. Machine learning can make connections that humans can't conceive of or even explain why they occur. And they can do it faster and at an extreme scale. Netflix's aim, for example, is to rank the best titles for each user so they can get to the content that they want faster. Netflix says it strives to rank titles in an order that the readers will most enjoy and with minimal effort.[1] People can be shown relevant content they didn't even know they wanted, made possible through machine learning. While that's a benefit to the content companies, it's a double-edged sword for the viewer inclined to binge-watch. Also, the recommendations aren't perfect, and we will look into why.

Figuring out how this machine learning works helps understand why such connections occur and why companies are scrambling to adopt recommendation engines. Further, recommenders are now crucial to modern companies with digital-first strategies in order for them to stay competitive and become or remain successful. As technology morphs into the backbone of more and more companies, gaining an edge is key. Here's *Harvard Business Review (HBR)*: "Recommenders' true genius comes from their opportunity to build virtuous business cycles: The more people use them, the more valuable they become; the more valuable they become, the more people use them."[2]

These recommenders are now so important, Schrage argues in *HBR*, that legacy companies risk being left behind in the digital age if they don't adopt them. Corporations that view recommenders as simply a marketing gimmick miss out on crucial digital information interaction with their customers. The fact that these systems continuously learn makes them constantly valuable to the companies that employ them. The most important distinction between digitally native and legacy companies isn't staff, datasets, or computing power but committing to deliver customer recommendations, he says. Companies are increasingly putting machine learning at the heart of their business.

Netflix's Famed System

At Netflix, content curation is based on information including other users with similar tastes, your past interactions, genre, categories, actors, release years, the time of day you watch, the devices you use, and how long you watch. Netflix plugs these data into its algorithms and ranks each title within each row, then ranks the rows themselves, to show the categories that it thinks the viewer would be most interested in. (It says it doesn't use demographic information, such as gender, among these data.) Netflix is among the most famous examples of machine learning for recommending media content, with about 233 million paid memberships as of the first three months of 2023.[3] Showing the power of recommendation, about 80% of the hours streamed at the company are influenced by choices from the recommender, according to Neil Hunt, Netflix's former chief product officer. It's not just about machines. Todd Yellin, Netflix's ex-vice president of product, explains the following:

> *In 2006, when I first got to Netflix and it was, how do we personalize? How do we do better putting the right title in front of the right person at the right time? And people always think, 'Oh, it's all algorithms.' No, no, no, no. This is a collaboration that's really driven by people because we put in the information.*[4]

Yellin goes on to say that Netflix set about categorizing its content by type. This is a major understatement. The upshot was an amazingly detailed categorization where Netflix broke down its content into more than 76,000 microgenres, according to Alexis Madrigal, who wrote a story for the *Atlantic* detailing the topic.[5] Such targeted slicing and dicing enabled Netflix to be hyper-specific in describing its own content and then targeting this content at its users:

> *Everything about a movie or TV show that I could think of, whether it took place in outer-space, it was a western in outer space and it had chimpanzees in it. And bam, you put that together and you have an outer-space chimpanzee western.*[6]

Netflix uses a system of secret category IDs to break this down. Some users have even listed them on the internet to use as a shortcut to find their favorite genres.[7] Some examples would be Scary Cult Movies From the 1980s (2), Understated Biographical Documentaries (16), and Exciting Film Noir (8).[8] What such a granular database allowed Netflix to do was precisely target movies and TV in a way tailored to users so that they would not only locate the films they want but also find similar shows that they didn't even know they wanted to watch. Adding machine learning on top of that let the company scale its tailored approach in a huge way, contributing to its success.

In Detail: Netflix Algorithms Unpacked

Netflix uses a whole variety of algorithms to wrangle its content. Among them are personalized video ranking, page generation, search, and similarity ratings. On the homepage, Netflix's algorithms use three levels of personalization for each row: the choice of row, which titles appear in that row, and the ranking of each individual title. The company employs A/B testing to see which techniques work best on an ongoing basis and is constantly striving to improve the effectiveness. Netflix acknowledges that predicting what consumers want to watch is hard because dealing with consumer tastes and preferences is "an extremely challenging problem."[9] Each user's choices are nuanced, and people's tastes change over time. The company recognizes that there's still a long way to go before a member gets the perfect recommendation at the top of the page. 'Trending Now' is a row of videos calculated in real time, as events happen, allowing Netflix to tailor content to events that affect movie-watching, such as Christmas, Halloween, or the Oscars. Netflix collects two streams of data: videos played by members and videos seen in the viewports. The Trending Now row is computed using live user data, aggregated play popularity, and other signals like members' viewing history and past ratings.

Netflix uses unsupervised learning methods, such as clustering algorithms, as well as several supervised classifiers. The company lists at least 12 techniques in use, including linear regression, logistic regression, and elastic nets. The ultimate aim of the recommender is "optimizing the probability a member chooses to watch a title and enjoys it enough to come back to the service," the company says.[10] All of Netflix's machine-learning technology is geared toward this. Next, Netflix is shifting to auto-playing content it thinks viewers want to avoid the indecision of scrolling. That led to the implementation of the "play something" suggestion, with a shuffle sign, at login. That takes recommendations to the next level and avoids the user having to make any choice.

Another peek inside the black box reveals the extent to which Netflix tests and tweaks its algorithms in minute detail to improve performance. One aspect the company looked at was which images to attach to a variety of different shows. They studied the impact and effectiveness of a variety of images and tested them against one another to ensure they improved over time. Pictures have a strong impact in determining how users choose, in part because such decisions are made lightning fast by viewers, giving Netflix a short window of opportunity to capture member interest. This also makes the user interface of the home screen hyper-important and, by extension, heightens the importance of the algorithm's selections for users' choice. There's only a small amount of space on the front page, so the algorithmic choices are key. Images with a dragon breathing fire performed "significantly" better than other ads for *Dragons: Race to the Edge*, in testing. Netflix engineers concluded that using visible recognizable and polarizing characters was the best tactic for engagement. Also, images with strong facial expressions tended to do well. The image that performed the best in promoting the TV show *Unbreakable Kimmy Schmidt* had faces of two characters with clear, comical expressions. Netflix is also using machine learning more broadly to shape its catalog to gauge characteristics of what makes content successful. Algorithms also direct advertising spending and channel mixes and help optimize original content, according to Netflix's own machine-learning blog.

Everything at Amazon: Books, Products, TV, and Music

Amazon has been using recommenders for more than 20 years. It made a revolutionary discovery early on in its efforts and made a subsequent decision. That was to focus on the items recommended rather than the user. Amazon's aim is to build a store unique to each customer, tailored to their needs, and which looks different from anyone else's profile. In the 1990s, algorithms tended to be user-based, meaning they relied on data from users. Amazon broke with this to base its algorithm on the items. So instead of focusing on user data, and what similar-profiled users might buy, Amazon looked at the item itself and created a way of finding a similar relevant product that the user would be tempted to buy along with it:

> *The better way was to base product recommendations not on similarities between customers but on correlations between products. Amazon's Personalization team found, empirically, that analyzing purchase histories at the item level yielded better recommendations than analyzing them at the customer level.*[11]

As well as improving the quality of recommendations, it had huge advantages in computing resources. Looking up the user history and comparing it across the universe of Amazon's customers would be a huge task, whereas comparing a product to relevant products is faster and requires less firepower. This technique gave rise to a recommender that spread across the internet. Amazon got its start in books. So cataloging was natural and essential. Recommending was one step from that. Now,

Amazon's recommender is one of the most powerful in the world. And its engineers are constantly thinking about the future:

> This moves beyond the current paradigm of typing search keywords in a box and navigating a website. Instead, discovery should be like talking with a friend who knows you, knows what you like, works with you at every step, and anticipates your needs. This is a vision where intelligence is everywhere. Every interaction should reflect who you are and what you like, and help you find what other people like you have already discovered.[12]

Amazon likes to highlight key products on its home page, and being listed there is likely to give a product an incredible sales boost.

In Detail: Amazon Algorithms Unpacked

Like Netflix, Amazon also conducts extensive A/B testing to see which algorithms work the best to achieve its aims. Recommender systems are often considered incomplete matrices with a series of blanks. Amazon decided to use a machine-learning technology called deep neural networks to help fill in the missing data of the matrix for its Prime Video recommender. The matrix completion typically uses a type of unsupervised neural network called an auto-encoder. These take a set of inputs and compress them and encode them and then reconstruct them in order to filter out noisy data. Amazon researchers found auto-encoders performed worse than an item-to-item collaborative-filtering algorithm and even worse than a simple best-seller list. So Amazon's team took the auto-encoder and trained it on movie information sorted chronologically. This worked better. The upshot was that Amazon researchers' work led to a twofold improvement in that algorithm's performance, which Jeff Wilke, the CEO of Amazon's consumer division at the time, described as a "once-in-a-decade leap."[13] Since Prime Video's interface displayed six movies on the page associated with each title in its catalog, the researchers measured the system on whether at least one of its top six recommendations was in fact a movie that the customer watched in the following two weeks. This auto-encoder outperformed the bestseller list, and it also beat item-to-item collaborative filtering. As Wilke put it, "We had a winner."

Delving Under the Hood: Technical Details

There are three broad types of recommender systems: collaborative filtering, content-based filtering, and a hybrid model, which mixes the two. Collaborative filtering collects and analyzes user behavior and preferences to predict what a user would like. Content filtering looks at the similarity of items based on customer preference. Hybrid systems collect both collaborative metadata (basically tagging and labeling) and content-based transactional data. One common problem with such systems that rely on users' past preferences is how to handle new users who have no history. This is known as the "cold start problem."[14] This is one reason that when you sign up for a streaming service, such as HBO Max or Hulu, they immediately ask you to click on examples of movies and TV you like, so they can improve recommendations from the get-go. Once the user is established, engines rely on relationships between customers and products. Three main linkages to consider are user to product, user to user, and product to product. These can be seen in the Netflix and Amazon examples.

Apart from relationships, various other data are needed, specifically user behavior data, demographic data, and product attribute data. User interaction data can be explicit — the customer clicked "like" — or implicit — the customer watched a show five times. Product attribute data consist of genres for books or cast for films, for instance. There is also the problem of when the recommenders fail, by annoying the user, providing a flood of content in which they have no interest or which even actively repels them from the service. As Netflix acknowledges, the whims of users' moods and personal tastes are hard problems to solve. It's a work in progress as the technology improves and data volumes surge. There are arguments that recommenders sap our decision-making capabilities. Also, recommenders in social media can be troublesome for different reasons, including fake news and false connections.

It would be a mistake to confine recommendation systems to e-commerce.[15] They provide content, advice, and help with decisions. They can be useful and reusable anywhere suggestions are needed. Schrage in *HBR* suggests using machine learning for hiring product people or business partners and says wherever there's digitization and data, the power of recommenders will follow.[16] As we'll see in future chapters, the mechanisms behind these recommender systems drive other technologies spreading throughout the industry, from healthcare, to retail, to finance. But the meshing of domain knowledge with computer science knowhow is something more and more companies are grappling with as they embrace machine learning. The further they are from Silicon Valley, the harder they find it to get the staff and the expertise they need. Next, we move to something that contains machine learning at its heart: the search engine.

Chapter 3

Search and Seek

Search engines are machine-learning mammoths. They require massive computing firepower and house inordinate amounts of data as they scour the web for answers to our questions. Here, we explore search dynamics and how machine learning powers answers, rankings, and recommendations. Google's search and recommendation engine is a whole category of its own, which warrants special attention. We also examine the outlook for the Microsoft search engine Bing, which is integrating generative AI, and other search engine newcomers as the technology evolves.

Google Search is the ultimate machine learner. Alphabet's Google uses a slew of algorithms to find and rank queries. If you wonder how Google Search manages to pinpoint the answers you are searching for, how its searches improve over time, or why on occasion your searches miss the mark, there are a few key methods it uses underpinned by learning models. Each Google search looks at the words in a query, the relevance and usability of web pages, the source expertise, and user location and settings. Then it weights each factor according to the type of query entered.[1] More broadly, search engines use machine learning for pattern detection to help stop spam or duplication, identify rankings or signals, adjust weightings, interpret customized searches, and handle natural language. Machine learning is also used in image search, in ad quality, for synonyms, and to classify queries.[2] Further, deep learning, which uses neural networks with multiple layers, has enabled search engines to vastly improve their results over time.

Google's autocomplete function is where recommendations come into play, although in a very different way from Netflix or Amazon. When you type in a search, Google suggests words for you. Gmail offers to finish your email sentences using a similar setup. The way Google's algorithm considers any web search consists of five factors: the meaning behind the query, the relevancy of web pages, the quality of the content, the usability of pages, and the context. Google CEO Sundar Pichai says that artificial intelligence is impacting all of Google's products because it's baked into so many of its systems. He also says that AI is one of the most important technological advances, and while the concept is quite abstract, the more it's embedded, the more it becomes relied upon.[3] Google has designated itself as a "machine learning first" company for more than five years.[4] It is certainly at the forefront of technology, but as shall be seen, many other companies are advancing in this arena too.

Prediction and Rank

When you type a search, Google's autocomplete looks at the language, the location, the trending interest of the query, and your past searches. It can predict specific words as well as full search predictions. A system called RankBrain is Google's attempt to figure out the user's likely intent in making a search. This is a harder problem than it might appear. If RankBrain receives a phrase it hasn't seen before, it can guess at similar phrases with similar meanings using this filter to rank the results. The first place Google prediction goes is to the common or trending queries, says Danny Sullivan, Google's Public Liaison for Search: "For instance, if you were to type in 'best star trek...', we'd look for the common completions that would follow, such as 'best star trek series' or 'best star trek episodes.'"[5] And Google autocomplete strives to predict what users want rather than suggest a different search that they might like. This contrasts with other recommender systems, such as for videos or movies. But it's in line with a web search. Google admits that autocomplete searches aren't perfect and has deliberate policies in place to prevent dangerous or violent content appearing. Some predictions may still shock or surprise users.

Google is constantly hunting ways to discover intent within search queries. Its Language Model for Dialogue Applications (LaMDA) technology uses learning models to better understand conversation. Like other large language models, it's built on what's known as a transformer, so it can read many words and figure out their relationship to one another. But LaMDA was trained on dialogue, making it better at handling the meandering nature of conversation than most chatbots.[6] Google is mulling how to combine its search bar with such conversational interfaces and might make a product as an adjunct to search or to use in combination.[7]

In one more sign of the future direction of how machine learning will be adopted, Google is selling its Recommendations AI technology to retailers or other companies who don't want to build their own complex algorithmic recommender. The company based the product on its own experience in building machine-learning recommenders across its own products, like search, ads, and YouTube. For YouTube, the content is more like Netflix and Amazon videos. And similarly, recommendations are key to viewership. "Recommendations drive a significant amount of the overall viewership on YouTube, even more than channel subscriptions or search."[8] YouTube's system works using 80 billion pieces of information that it calls "signals." These include clicks on a video, watch time, sharing, likes, and dislikes.

Bing-o

Microsoft's search engine, Bing, has struggled to compete with Google and capture market share. But Bing has made considerable strides of its own in its use of machine learning systems as Microsoft pivoted toward artificial intelligence, and the technology is no less impressive. Bing is an example of AI at scale in that it highlights some of Microsoft's best and cutting-edge technology. Microsoft wants very large models and is building the infrastructure needed to train and deploy them for other organizations to use.[9] Bing, meanwhile, has been using large neural network models such as MT-DNN, Unicoder, and UniLM as Microsoft tries to improve search for its users. Before that, Bing was harnessing deep learning to make its search results smarter.[10]

Traditionally, search engines used keyword matching to match queries with results. Bing moved to create a number, known as a vector, for each word in an effort to determine meaning. Words with similar vectors have similar meanings and are clustered together in two dimensions. This deep-learning approach let Microsoft go further than keyword matching to gain better semantic understanding of search queries.[11] Microsoft says that using its Turing model for natural language understanding led to better caption generation and improved answers to search questions by as much as 125%.[12] One further improvement Microsoft focused on was to provide clear answers to yes or no questions, rather than providing a list of hopefully relevant links, as in the past. The system provides a yes/no answer through reasoning, after checking a variety of sources. Microsoft uses the example of "Can dogs eat chocolate?" An earlier Bing search example provided a list of factoids about chocolate. The new version clearly answers the following: No, chocolate is toxic to dogs. Search is improving with advances in Natural Language Processing, which we will look at in more detail in our chapter on text mining. But meaning and understanding are hard problems for search engines to handle. Search engines are further evolving with the use of something called transformers, models that examine words in relation to all other words in a sentence. Transformers take forward the previous technology in search by allowing words to be examined in context and all at once, rather than in sequence. For instance, Bing is harnessing ChatGPT technology from OpenAI to combine search with the tools of generative AI.

BERT

Google says that using its Bidirectional Encoder Representations from Transformers, known as BERT, does a better job of providing useful information in searches after a breakthrough in researching transformers. BERT can consider a word in the context of words that come before and after it in the query. It also uses the same technique to help achieve results across different languages.[13] So artificial intelligence, deep learning, and neural networks are all making search easier for users. Beforehand, search engines relied on algebra and formulas, along with matrix multiplication to line up queries to documents.[14] Today, as one example, search engines use neural networks to summarize text and figure out the similarities of words between documents. This allows for a vast improvement in performance. In search engines, using neural networks is known as neural information retrieval. The use of these networks lets search engines provide more relevant answers because of their ability to capture semantically similar words, generate text, provide translations or answers in other languages, as well as reduce searches that can't produce results.[15] This is key, as failing to return a search result will immediately send the user to another platform. Recommenders and search engines certainly aren't perfect. If you accidentally click on a movie you don't like, you risk being flooded with similar content. Your feed can clog with foreign content you may not understand. When shopping, if you buy an item, you are often bombarded with similar suggestions, even though once you've bought a thing you have no reason to repeat the purchase immediately. Buying a bed causes a stream of superfluous bed ads, for example. Fixing this ubiquitous problem must be in the works at sites like Amazon. Search engines still throw out some wacky and useless results. On the whole, they have improved and are improving all the time; indeed they are an indispensable part of modern life — the interface between the digital world and the real world.

Search has been an exploding field for careers in the past decade. So much of modern life is about databases, and queries to those databases, that the engineers and product managers who can figure out the answers to these problems are instantly in demand. Improvements in machine learning are accelerating this. Jobs related to search have been expanding to account for both search engine optimization, as companies seek to improve their results and rankings within search engines, and technology companies with the systems behind search itself. There's a shortage of machine learning engineers with expertise in recommenders, specifically for those who know about the infrastructure of such systems.[16] The future of search is likely to be in ever-better information retrieval, along with better user interfaces, such as predictive queries to guess what you want to ask or voice recognition to let you ask questions orally rather than type them into your search box. It's also likely to combine the power of large-language models used by the latest chatbots along with voice recognition and the latest question-answer techniques. We'll look at this in our chapters on virtual assistants and the Internet of Things. First, we turn to a key theme, that of computer vision.

Chapter 4

Cool Cats and Hotdogs

In this chapter, we explore how computer vision works and how machines first translate pictures into pixels and then into zeros and ones. We go on to study how data are used to inform the identification of images and faces. Then, going back in time to 1989, the first handwriting detection systems used to read bank checks are investigated, followed by an analysis of how machines learned to identify cats and why that was such a technological watershed.

In 2012, an event kicked off the current arms race in deep learning and computer vision. It centered on computers teaching themselves to find cats on the internet. Academic researcher Quoc Le was frustrated with machine-learning systems that could only detect images from data labeled by people. Images of hotdogs, say, labeled as hotdogs, or burgers labeled as burgers, and so on. Here, the machine is fed the labeled data and learns what a hotdog looks like through pattern recognition if enough images are presented. Le wanted to cut down on the work of labeling and have the computers figure out for themselves how to identify images. This is called unsupervised learning in programming parlance, as described in the first chapter. Images are fed to the algorithm without labels, and the machine looks for patterns and commonalities. This can yield some surprising results. And can be mysteriously effective.

Computer vision is just one application of unsupervised learning, but it's a key one. It's essentially how machines learn to see. IBM defines it as enabling "computers and systems to derive meaningful information from digital images, videos and other visual inputs — and take actions or make recommendations based on that information."[1] Image recognition has become enmeshed in our lives, from unlocking your phone just from a face and without clumsy passwords, to discovering online products to buy. Photographs can automatically tag themselves to people on social networks, and criminals can be caught on video and identified. But in 2012, such technology was still unproven. Le and a group of Google researchers hooked up 16,000 computer processors and downloaded 10 million images from YouTube to create a neural network of 1 billion connections and then set their algorithm loose. The experiment was new in that this was one of the biggest neural networks built to date. And people didn't think it would work. But it did. The machine learned to recognize cat faces without being told anything about cats. This was a seminal moment in machine learning.

Dog or Cat?

Those 10 million images were selected randomly from YouTube videos. The researchers didn't tell the machine what a cat was, or to look for it, but the algorithm found them anyway, figuring out for itself what cats look like. The sheer amount of data meant that the patterns and similarities between the pictures became clearer. Le, his Stanford University supervisor Andrew Ng, and the other researchers published a paper with their findings. The experiment also improved the accuracy of classifying images generally and proved unsupervised learning could improve the accuracy of machine-learning systems. The scale of the operation led to learning systems that were previously impossible, *The New York Times* reported at the time.[2] The Google experiment also improved accuracy in image detection by 70%. "This is something we're really focused on — how to develop machine learning systems that scale well, so that we can take advantage of vast sets of unlabeled training data," the Google researchers said on their blog.[3] "This network had never been told what a cat was, nor was it given even a single image labeled as a cat."

Further, this network that taught itself to identify cats seems to mirror the way neurons work in the human brain when identifying objects.

How Exactly Does Computer Vision Work?

Computer vision doesn't have anything magical or mysterious about it. When you see the amazing applications of facial recognition by Facebook or Apple products, it might seem incredible, but in the end it's all just pixels reduced to numbers. Machines' narrow focus is summed up by a scene on the technology spoof comedy TV show *Silicon Valley*, where the character Jian-Yang builds an app that distinguishes hotdogs snapped on a cellphone camera. The housemates are amazed at the app's brilliance until he tries to use it to identify pizza. Unfortunately, the app only detects two classes of objects: "hotdog" and "not hotdog," so pizza is only identified as "not hotdog." The joke is real, however. Image recognition has a narrow scope. This is one reason self-driving cars are so hard to build. As we shall see in a later chapter, images in three dimensions are even more difficult for software to perceive than pictures.

Not a Hotdog

In computer vision, the two main tasks are classifying objects and identifying objects. When a picture is rendered for a computer, it just looks at a grid of pixels. Each pixel corresponds to numbers representing red, green, or blue light. So all computers see in an image is a grid of numbers, from 0 to 255. (The total number relates to the quantity of pixels you need to represent red, blue, and green: 2 to the power of 8.) Then, a machine-learning model takes a series of pictures and looks at similarities and patterns, across, say, 10 million such grids of numbers. Specifically, it detects the edge of shapes. Think of how people solve jigsaw puzzles by finding the edges of objects within the jigsaw's picture. Sorting pieces into sky or water and trees makes it easier to reconstruct the whole jigsaw. Machines need context to identify objects, the same as humans. The models add this context. In computer vision, a method called Convolutional Neural Networks (CNNs) is often used. This works by breaking the image down into pieces, running different "layers" over each piece, making predictions against the labels of the image, and then iterating to improve. The larger the amount of labeled data, the better the machine's predictions. Layers represent part of the structure of a machine-learning network.

To a computer, a dog and a cat are not innately different, except for a divergence in these number patterns, showing whisker length, nose size, eye shape, and so forth. So a doglike cat or a feline-looking dog might trip up the machine, which has no concept of any inherent difference. But it might have seen enough dogs that it can detect even the subtlest differences between dogs and cats. But it's still just pixels. Computers don't understand what dogs or cats are nor can they gauge the difference in any human sense.

Yann's Handwriting

Go back to 1989, and Yann LeCun, a computer science academic who went on to become Facebook's chief AI scientist and a legend in machine learning, figured out a way to get computers to read handwriting. This optical character recognition system was used by banks to read checks and zip codes. LeCun combined CNNs with something called backpropagation. Again, this was just pixels, but LeCun's method handled the uncertainty and variability of handwritten letters. And the techniques were the foundation of machine learning in later years. LeCun went on to become one of deep learning's key architects.

Recaptcha and Captcha

In the year 2000, Yahoo! faced a problem. Spammers were creating fake email accounts on the service. Millions of them. The spammers built bots that created these accounts automatically, and Yahoo! didn't know what to do about it. Luis Von Ahn was a graduate student at Carnegie Mellon University, and he was interested in devising a test that could tell computers and humans apart. He called it the Completely Automated Public Turing test to tell Computers and Humans Apart, or Captcha. After showing it to Yahoo!'s chief scientist, the email and search company embedded it in their registration process within weeks. The Captcha puzzles were little words or numbers, made of pictures which computers at the time couldn't read. The image would distort the text such that robots wouldn't be able to tell what they were, while people would instinctively see them. Van Ahn wasn't the only researcher to develop these kinds of tests. The first patent was in 1997, though it didn't use the term Captcha. Soon, though, some 200 million such tests were in use on the internet. Trouble was, people couldn't always read the Captchas either, and they hated the seconds or minutes wasted trying to decipher them. Von Ahn wanted to update the test. Meanwhile, during the 2000s, there was a big push to digitize books and other offline writings. Old books and documents could be scanned onto computers well enough, but machines still couldn't read the letters or turn them into searchable text. Then, Von Ahn had a brainwave. He could get computers to learn to read scanned letters at the same time as people typed in Captcha tests. Halving the work while doubling the output — and for free. The elegance of enhancing internet security while teaching the machines was irresistible. "It occurred to me that you could take all of the words that the computer could not recognize, and we could get people to read them for us while they were typing Captchas on the Internet," Von Ahn said in an interview with National Public Radio.[4]

The New York Times wanted to digitize its 130-year-old archives, and after watching a talk by Von Ahn, the newspaper's chief information officer hired him to use the technology to move its back catalog online. Von Ahn rebranded Captcha as Recaptcha and went from giving it away for free to selling the technology. *The New York Times* was the first client. The way it worked was as follows: When you solved a Captcha, you were given a picture of a word from the Times that computers couldn't read. As the users solved the puzzle, they also taught the computer what those scanned letters were. In 2009, Google bought the updated Recaptcha to help it figure out how to computerize the world's text. It has managed to scan at least 25 million books from university libraries. Von Ahn went on to found the language-learning website *Duolingo*, another company at the forefront of machine learning but in language education.

ImageNet: Datasets Versus Better Algorithms

The next computer-vision breakthrough came in 2006 when computer science professor Fei-Fei Li was among the first to recognize the value of better datasets rather than better algorithms. She created ImageNet, a giant dataset of images that evolved into a competition, to see whose algorithm worked best at identifying objects in the pictures, with the lowest error rate. It was the start of a big realization. Algorithms improved when trained with the ImageNet data. A further feat in cat identification occurred in 2013, with another competition, this one from Kaggle, to make the most accurate detector of cats over dogs in 25,000 photos. Researcher Pierre Sermanet won the competition, using a CNN, with results of 98.9% accuracy.

In Detail: Convolutional Neural Networks Unpacked

CNNs are a type of deep-learning algorithm that extract features from images. These networks take in a two-dimensional image and then assign importance to aspects of the image. Then they differentiate the different objects within the image, giving out various weightings as levels of importance. The CNN is so-called because it is seen as analogous to the architecture of neurons in the human brain and is inspired by the visual cortex. Central to the operation is the convolutional layer in the network, which performs a "convolution" as it changes the weights it gives to the importance of each of the inputs. The object of convolutions is to detect features in the image, like edges or eyes. Machine Learning Mastery describes it as follows: "In summary, we have an input, such as an image of pixel values, and we have a filter, which is a set of weights, and the filter is systematically applied to the input data to create a feature map."[5]

The convolution is just combining two functions to create a third function, multiplying the weights given to each feature. Convolutional Neural Networks learn from the input images by altering their weightings and biases in order to improve predictions. They typically have an input layer, an output layer, and multiple hidden layers in between, as well as convolutional layers that contain other layers that perform other functions, like pooling, normalization, and rectified linear activation.[6] So CNNs became key in computer vision and facial recognition because of their accuracy. While they were invented in the 1980s, they weren't usable until the leap in data and computing power enabled by Graphics Processing Units, or GPUs, two decades later. Now there is power and data enough for them to be feasibly used.

I See Your Face

If it seems scary that computer vision is improving at such speed, that's because it could be. Microsoft President Brad Smith is among those in the industry highlighting these dangers, specifically of facial recognition, as we'll see in detail in the following chapter. First, there's the risk of bias if companies can detect your name from your face. Second, widespread adoption can intrude on people's privacy. Third, mass surveillance by a government can encroach on societal freedoms. It's illuminating that Microsoft, which has some of the most advanced facial recognition software in the world, is warning of these dangers and even pressing for government regulation on the topic. China and Russia are adopting the technology, as are governments of Japan, the US, and Israel, among others.

There are also accusations of bias in facial recognition, which we will explore further in later chapters. That bias occurs because the datasets of photos used to train the models didn't include enough faces from different places across the world, meaning that the software was better at identifying white faces. Computer vision is getting better and more accurate all the time. Google, Amazon, and Microsoft all have image recognition technology that's almost on a par with human hand-tagged images. Giant databases of billions of photos of faces from Facebook's Instagram, Smugmug's Flickr, and the rest, combined with the increased processing power of the latest chips, mean ever-increasing accuracy at identifying people from their facial features. Facebook's facial recognition technology in particular has generated controversy, especially around privacy, partly because it's so accurate.

Facebook in November 2021 announced it would shutter its facial recognition system and delete more than 1 billion people from its database of templates that pick out individual faces. The company cited "societal concerns" and said regulators haven't provided clear rules.[7] Kai-Fu Lee, the famed AI venture capitalist who formerly worked for Microsoft and Google, says that in China facial recognition is now being used to discern emotions and estimate age. Similar algorithms are now being used for something even more complicated: language. In later chapters, we'll see how robots scan text at speed, mine words for meaning, and even write whole sentences and paragraphs. But first, we will unpack some of the privacy concerns around letting computers identify faces.

Chapter 5

Smile for the Camera

Facial recognition technology is spreading. In these pages, we examine why, what's under the hood, and how allowing machines to automatically identify your face has a dangerous side. Facial recognition is one of the main areas in artificial intelligence where ethics and technology collide. Companies are getting more wary about deploying it, and people are becoming more wary of how their biometrics are being captured. We also look at how China is using vision technology.

Computers can now pick your face out of a photograph or some video. Your image can even be captured without your knowledge and stored online. A kind of biometrics, facial recognition has improved by leaps and bounds in recent years and the technological advance is accelerating. As it does, it makes consumer products better and digital security tighter, but the areas of ethical concern skyrocket. Facial recognition is built on computer vision. The way it works is that computers detect a face from an image and then match that face to a picture in a database. To do this, the machine must perceive that a face exists within the frame, handle the different alignments or angles, recognize that face, and verify the face against a person's identity.

Technology companies are using facial recognition to let people quickly and conveniently unlock their phones. Law enforcement agencies are using it to help identify and catch criminals. Airport security uses smart cameras to check identities and scan crowds, averting possible terrorism and making flying safer. Companies are using systems that recognize faces in all kinds of ways for marketing their products. Facial recognition is one of the areas of machine learning where the results can appear truly amazing. That computers can locate faces in pictures, identify them, and verify the person against a database is no mean feat. Recognition software "reads the geometry of your face."[1] That means mapping various distances between facial features, like your eyes, nose, mouth, and chin. Then from that, it creates a facial signature in the form of numbers. That signature formula is then matched against a database of known faces to determine your identity. It's unique in a similar way to fingerprints.

Dangers Abound

The dangers of deploying this technology at scale are all too clear. The power that the latest facial recognition technology imbues on governments to track or monitor their populations is unprecedented. The technology can be misused, and basic freedoms risk being eroded as it inevitably proliferates. Then there are privacy issues regarding your personal data being stored across multiple databases. Security problems arise if these data are stolen. There's also the chance facial recognition software makes a mistake and mixes up an innocent bystander with a criminal, for example. Then there's the risk of bias in the original model, making certain ethnic groups harder to recognize or categorize because of a dearth of data. Finally, facial recognition is likely to spread everywhere across society, making anonymity harder. Some countries place a higher value on privacy than others. These issues are already raising alarm from many quarters. Balancing the usefulness of this technology against the dangers it poses is something that companies and governments are grappling with. And several companies have decided the risks are too much. There's also a growing push to regulate the use of facial recognition in certain countries and jurisdictions.[2]

Face to Facial

Amazon's Rekognition software is one of the most advanced in the industry.[3] It can tell you where faces occur in photos or video, facial "landmarks" such as the position of the eyes, and even whether there are emotions, like happiness or sadness, contained within the picture. Rekognition provides users with a confidence score for the face and facial attributes.[4] Facebook has a technology called Deep Face, which uses a neural network to analyze 68 data points on a photograph to measure facial features, proportions, and color.[5] When you upload a photo to the platform, it can jump to suggest the identity of the person.[6] However, Facebook parent Meta has expressed caution in how its technology is used in 2021, shutting off certain automatic facial recognition functions on its social network.

I See Your Face

San Francisco voted to ban facial recognition usage by authorities in 2019, amid concerns about privacy. Boston and Oakland have also banned its use by government.[7] In Britain, a campaign against facial recognition is striving to block mass surveillance in public places that use cameras which pick out individual faces. Police and private companies have used the technology in malls, concerts, bars, museums, and stadiums. There's no British law regarding facial recognition and more than 3,000 people have been wrongly identified by police recognition cameras.[8] Microsoft, which also has powerful facial recognition software, was one of the first companies to recognize the dangers of the concept in practice. The company began advocating for laws to regulate its safe use, an unusual stance for a technology company pushing for regulation of its own products.[9] Microsoft President Brad Smith identified and highlighted three problems back in 2018: bias and discrimination, privacy intrusions, and governmental mass surveillance harming democratic freedoms.[10] "Don't ask what computers can do, ask what they should do," says Smith. Microsoft raised the alarm about the dangers of facial recognition early compared with other practitioners and went on to adopt six facial recognition principles to use as a foundation when building out the technology. They are fairness, transparency, accountability, non-discrimination, notice/consent, and lawful surveillance.[11]

In terms of facial recognition, better sensors, cameras, and deep learning techniques combined with the explosion of datasets of faces available online make it possible to identify people on an unprecedented scale and in real time. That makes the need for government regulation all the more urgent unless you want to all-but-eliminate privacy anywhere there's a camera. And Microsoft also sees the need for international cooperation. The uses of facial recognition aren't all malign. It can be used to find missing children, for example, or to identify and capture criminals. But it can also be used to monitor the behavior of dissidents or radicals and as a tool of suppression and control. More recently, Microsoft said it would phase out software that detects emotions, as well as remove face recognition features that can pinpoint gender, age, facial hair, smiles, or makeup. These features are ripe for bias and abuse. The reason the company gave was its aim of adopting responsible AI standards.[12] Other companies have followed suit in expressing caution. In November 2021, Meta announced it would shut down its facial recognition system on Facebook and that it would delete more than 1 billion face recognition templates. That the company built 1 billion templates speaks to the power and range of facial recognition and also of the lengths companies will go to achieve reach and control.[13] Meta said the useful instances of the technology must be weighed against societal concerns about facial recognition as a whole. Meta also said with regulation at an early stage, it's better to limit the use of facial recognition, as well as openly debate its use, including those most affected by it.

Police Ban

Amazon banned law enforcement from using its Rekognition software because of concerns about racial bias and false arrests.[14] So as we can see, the dangers are in some cases outweighing the benefits of facial recognition, even to the large technology companies that build and sell such services. Data on faces are easy for authorities to collect and hard for the public to avoid yielding up.[15] Also, the data can be shared across agencies, while cameras are getting more powerful. Along with autonomous weapons, facial recognition has the ability to change society in a large and unforeseen way, and the laws and regulations haven't kept pace with the innovation. Companies that can control the usage of facial recognition data in a granular and safe way will be poised to benefit from the increased scrutiny of how these systems are used.

China has the world's biggest facial recognition system. As part of its overall strategy on artificial intelligence, the government is transparent about its plans to monitor its citizens. The Chinese government has a different attitude to Western governments in the surveillance of its people, and the population's tolerance of being monitored is higher.[16] China in 2018 claimed it can scan its citizens in one second to fight crime and tighten security through a facial recognition network that's in place in 16 cities and provinces.[17] Tencent's WeChat app, which acts like a Chinese version of Facebook, WhatsApp, Uber, and PayPal all rolled into one, uses facial recognition software that is widely in use in the country. Tencent is also one of the top applicants for patents on facial recognition in China. It used face recognition to monitor and limit screen time for under-18s, amid concern over youth gaming addiction.[18] Facial recognition software is used by the authorities but has also been in use commercially for years. Now the government is centralizing control of facial recognition, taking back control from companies.[19] There's also a backlash brewing as the government cracked down on the power of technology companies.[20] Ride-sharing giant Didi Global was fined by the Chinese government in 2022 over its data security laws, including misuse of facial recognition data.[21]

Blind Help

What about the benefits of image detection? Both Facebook and Microsoft have accessibility software for blind people that harness facial recognition features to help users navigate the world around them. For law enforcement, facial recognition certainly makes it easier to pinpoint and catch criminals, and digital security is harder to hack with facial recognition. Missing persons can be identified and found, and shopping can be made more efficient.[22] It's also widely used for security and anti-fraud in the gambling industry and in banking. British bank HSBC has an app using facial recognition to log in to its banking service, for example. Casinos have used it to detect banned customers and known gambling addicts.

Technology Details: How It Works

First, faces need to be detected and positioned within the image. Then, pattern recognition captures the facial structure. Images are normalized and then mapped against a gallery of faces.[23] In his book on facial recognition, Ian Berle describes this process as acquisition and pre-processing, feature extraction, classification, and identification/verification. There are many different algorithms to achieve it. In terms of the pros, facial recognition makes flying safer. Airports use it, along with the Department of Homeland Security. It's more efficient than passwords to unlock phones and bank accounts. Camera technology can help stores catch criminals, simplify security checks both by authorities and on personal devices, and help find missing persons. And for the cons, it's a threat to privacy, mistaken identity is a risk, and the technology can be tricked.

The size of the global facial recognition market is predicted to grow to $13.8 billion by 2028 from $4.5 billion in 2020, according to Emergen Research.[24] So there's interest across industry, even with the doubts. The future of facial recognition is likely to reside with regulators, as the pace of technological change is outpacing what is socially acceptable in many places. But the scale that machine learning allows means that facial recognition surveillance would be possible on a mass scale. There's also the worry about deep fakes and the risk of bias. Once you start combining the various technologies, for example, by attaching cameras to drones, which can detect faces, the worries start to ratchet up. The military applications are where we turn next, with autonomous weapons.

Chapter 6

Drone Swarms and Slaughterbots

Robots and drones are now capable of deploying to the battlefield without human control. There are pros and cons of building them, which must be balanced with the inevitability of the technology. What are Lethal Autonomous Weapons, how do they work, and what ethical concerns do they raise? The United Nations is holding state-level discussions on autonomous weapons and their use. In this chapter, we examine how machine learning is being applied in the defense industry, as well as the state of the art in military hardware and software, which is considered yet another ethical minefield.

Killer robots aren't what you think. Neither are they what's portrayed in the movies. In real life, they are at once more mundane and more terrifying. One of the most controversial and thorny areas in artificial intelligence, these machines, capable of acting without human control or oversight, are described by the United Nations as Lethal Autonomous Weapons Systems, or LAWS. Arms-control advocates use the livelier term "slaughterbots." The idea that robots can make decisions on firing weapons in life or death situations is universally distasteful. But the technology is already here, leaving it a matter of deployment, scale, and degree. And proponents say that autonomous weapons have the capacity to be more accurate and faster than people, potentially reducing the loss of life to soldiers on the battlefield. The thought that robots will somehow gain consciousness and take over our defense systems like rogue computer Skynet in the movie *Terminator* is seen as far-fetched by technologists. The reality is far more prosaic but also deadly to combatants and civilians. Gary Marcus and Ernest Davis describe it well as follows:

> *The real risk is not superintelligence, it is idiots savants with power, such as autonomous weapons that could target people, with no values to constrain them. For now, we are in a kind of interregnum: Narrow but networked intelligences with autonomy, but too little genuine intelligence to be able to reason about the consequences of that power.*[1]

Robot Turrets

Examples of autonomous weapons include robot turrets used to repel missiles, drones set to seek certain targets using facial-recognition technology, and unmanned land or sea vehicles capable of carrying weapons. In a 2016 Dallas police shooting, officers in a standoff with a holed-up shooter ultimately deployed a retooled bomb-disposal robot to deliver a bomb to the gunman. It was the first time US police used a robot to kill a suspect.[2] The United Nations is starting to look into the issue of autonomous weapons, to reach state-wide consensus on what constitutes them and how they will be governed. UN Secretary-General António Guterres stated that "machines with the power and discretion to take lives without human involvement are politically unacceptable, morally repugnant and should be prohibited by international law."[3] Guterres sent a message of encouragement to a 2019 Geneva meeting of governmental experts, saying their work was of utmost importance. A group of 70 states issued a joint call for action on autonomous weapons at the United Nations General Assembly.[4] However, as of early 2023, no concrete progress has been made at the governmental level on banning or regulating the technologies.[5]

Machine Learning in Defense

So where exactly does machine learning fit into autonomous weapons? Machine learning is used in target detection, a form of image recognition. It can be used in anomaly detection, where pattern recognition is used to alert for outliers. It can also be used to manage information during reconnaissance. Drones can send streams of data back during surveillance, and machine learning can help analyze such data. Finally, there are so-called decision support systems, where results can help inform decisions based on data analysis.[6] While weapon research programs are investigating some of these uses, there is still a gap between "fielded weapons" and experimental tools.[7] But the US has been active in using the technology for a decade or more:

> *Automatically inducing knowledge from data has generally been found to be more effective than manual knowledge engineering. Development of state-of-the-art systems in computer vision, robotics, natural language processing and planning now rely extensively on automatic learning from training data.*[8]

As for the latest kinds of machine learning needed for weaponry, Distributed Reinforcement Learning is one example being explored to control multiple drones in order to avoid collisions and to expand the number of drones that can work in coordination at once, which we explore in the following when examining swarms. Companies, rather than countries, will be on the front line of developing such technologies and systems, and corporations including Microsoft and Google have become embroiled in the ethical dilemmas involved. Google quit the US Defense Department's Project Maven computer vision program after its employees protested, and Microsoft and Amazon picked up the sub-contracts. Traditional defense companies like Lockheed Martin and Boeing are already making deployable autonomous systems.

Gunpowder, Nukes

Experts say these weapons represent a third revolution in the history of warfare technology, after the discovery of gunpowder and the invention of nuclear arms, and will have as big an impact on the future of conflict.[9] As usual, definitions alone are a tricky enough problem. What constitutes autonomous in a military context, and which kind of systems count? What are the distinctions between automatic, automated, and autonomous, and where is the line separating weapons and technology? There is no internationally agreed-upon definition for what constitutes lethal autonomous weapons. In one effort at clear categorization, the US Defense Department calls them "weapon system[s] that, once activated, can select and engage targets without further intervention by a human operator."[10]

Humans in the Loop

The US also recognizes human-supervised and semi-autonomous systems, described as Human on the Loop and Human in the Loop. (On-the-loop weapons can target themselves, but soldiers can abort them; in-the-loop weapons still need troops to pull the trigger.) The first report of human death from a fully autonomous weapon occurred in Libya in March 2020. According to the UN, soldiers of the Government of National Accord (GNA) were battling troops of the Haftar Affiliated Forces (HAF). A rotary wing attack drone was used to kill retreating HAF forces with no operator control:

> *Logistics convoys and retreating HAF were subsequently hunted down and remotely engaged by the unmanned combat aerial vehicles or the lethal autonomous weapons systems such as the STM Kargu-2... and other loitering munitions. The lethal autonomous weapons systems were programmed to attack targets without requiring data connectivity between the operator and the munition: In effect, a true 'fire, forget and find' capability.[11]*

There's still ambiguity about the incident, how many were killed, and where. But the Turkish-made STM Kargu-2 is a 7-kilogram flying quadcopter. It's capable of autonomous targeting, can fly in swarms, and stays operational when communications are jammed. It also has facial recognition technology to target people.[12] The risk of error is a major problem for autonomous weapons. And problems could be sparked by a technical issue, such as confusion caused by light backgrounds, or the difference between training situations (on which the computer models and algorithms are based) and chaotic combat situations. Other unforeseen circumstances in the confusion of the battlefield could yield surprise results from weapons without a human in the loop. System breakdowns or malfunctions could also cause unexpected numbers of casualties, escalate armed conflict, or even harm civilians. Then, there are the ethical worries of full autonomy.

Drone Swarms

Drone swarms are one area of particular concern for arms-control advocates and military experts alike. Such swarms are a horde of tiny autonomous robots that are able to overwhelm an enemy, striking their target even if some or many are taken down. They are possible to deploy by air, water, and land. One small drone can wreak havoc with a bomb. Many could cause catastrophe on a battlefield. But they need to operate together in sync, and such swarms make it too hard for individual human operators to coordinate multiple drones at once, meaning machine learning is needed so they can cooperate and operate in harmony. But at such a level of autonomy, ethics becomes an issue because of accountability. Aylett and Vargas lay it out in their analysis of robotics:

> Levels of automation and technologies taken from robotics have been applied to weapons for some time. The US cruise missile has for many years acted autonomously... It combines dead reckoning, GPS information, and contour matching of the terrain it flies over. The issue that really worries experts is giving such devices the power to

autonomously select targets, as well. Imagine a swarm of killer drones permanently flying above a city, deciding whom to kill and then killing them.[13]

The US Defense Advanced Research Projects Agency (Darpa) is already testing swarms utilizing more than 300 robots. The agency says it achieved coordinated operations from two different contractors with virtual and real swarms working in tandem and allowed the use of special gear such as artificial reality kits, phones, and tablets to control them.[14] Darpa has a program called OFFSET, or Offensive Swam Enabled Technologies, to investigate new technologies to use tiny robots in swarms, combining autonomy and human control. The swarm tactics inform the swarm technologies, Darpa says. The British military has more than 200 projects ranging from autonomous ships to drone swarms, along with using machine learning for talent management and predictive maintenance.[15] Proliferation of autonomous weapons is another potential global problem. In contrast with nuclear or biological weapons, autonomous weapons are low cost and easy to come by. The Future of Life Institute lays it out clearly as follows:

Slaughterbots do not require costly or hard-to-obtain raw materials, making them extremely cheap to mass-produce. They're also safe to transport and hard to detect. Once significant military powers begin manufacturing, these weapons systems are bound to proliferate.[16]

The Future Already Happened

The technology to meld machine learning with lethal autonomous weapons is in progress, regardless of ethical considerations or state-level non-proliferation agreements. How states and companies handle the development will be crucial. Russia's Kalashnikov has made strides in developing weapons harnessing machine learning techniques, and China has a government-backed AI plan to increase spending and has made advances in autonomous ground vehicles.[17] As AI expert Kai-Fu Lee says, machines lack common sense and the human ability to reason: "No matter how much you train an autonomous-weapon system, the limitation on domain will keep it from fully understanding the consequences of its actions."[18]

Weapons will become nimbler, smarter, and deadlier at an alarming pace.

Chapter 7

Word Miners

In this chapter, we investigate how machines scan documents at speed, and what benefits and challenges this ability brings. We look back into the history of technological development to understand where and how the Natural Language Processing breakthrough occurred. Computers can't read (in any true sense), and we also explore why that is, along with details of what computers can do and how the technology works. In doing so, we gain an understanding of text extraction, Natural Language Processing, and the recent developments in chatbots, leading to the following question: What, exactly, is a GPT?

If computers are so smart, how come they can't read?
— Marcus and Davis[1]

How long would it take you to read 100,000 words? Robots can do it in a flash, and that has implications that are only just being realized. For work where documents need to be scanned, having a computer read thousands or millions of words is an amazing prospect. Add to that the ability to summarize text or extract key information and you have a powerful proposition. Yet language is a difficult problem in computer science. That's because of the nature of ambiguity, complexity, and meaning. Words can have different meanings, sentences can be laden with context, and irony or sarcasm can be impossible for computers to detect. There's also the problem that language constantly shifts and morphs, colloquialism and slang can be tough to track, and computers can't understand jokes.

There are three main forms of machine learning when it comes to text: Natural Language Processing (NLP), Natural Language Understanding (NLU), and Natural Language Generation (NLG). The latter two correspond roughly to reading and writing, although as we'll see, it's slightly more complicated than that. The general term Natural Language Processing, is used for automatic translation of foreign languages, as well as text summarization, personal assistants, predictive text, and sentiment analysis. NLU is a subset that pertains to the scanning of text to extract meaning. NLG, on the other hand, is the process of machine-generated writing. This language generation takes data points and turns them into plain English. NLP breaks down language into small pieces called tokens and tries to derive meaning from them by looking at relationships. The ultimate aim of these machine learning systems is to read and decipher human languages in a useful way. In other words, they let computers interpret human language.

Roots of Learning

As with much else in machine learning, NLP had its beginnings in the 1950s with Alan Turing, who wondered if machines could think. In the famous Turing Test, the Cambridge mathematician conjectured that if a computer can imitate human language to a sufficient degree, it could be considered capable of thought. From the 1950s to the 1990s, symbolic natural language processing was the norm, with computers using rules and symbols to attempt to understand language. This changed toward statistical NLP in the 1990s, as more and more data became available. In the 2010s, this progressed to using neural networks as computing power increased. In 2011, Apple's Siri was one of the first NLP virtual assistants to be used and adopted broadly by consumers.

State of the Art

OpenAI was founded in 2015 to promote safe artificial intelligence. Billionaire entrepreneur Elon Musk and Y-Combinator's Sam Altman were its initial co-chairs (although Musk left in 2018). It also boasted an array of scientists and engineers as founding members, along with venture capital luminaries to provide funding. The outfit was formed as a non-profit, with a mission of advancing digital intelligence in a way that would benefit humanity and a goal of artificial general intelligence, something not yet achieved outside science fiction: A safe artificial intelligence capable of a breadth of tasks on the level of humans. One real-world impact of OpenAI is its gigantic natural language model, called a Generative Pretrained Transformer (GPT). This performs an array of natural language tasks, such as questions and answers based on existing knowledge, summarization, grammar correction, or extracting content information. It's based on unsupervised learning models. OpenAI initially made this system available in a limited way, allowing access to this huge language library. It acts like an autocomplete function on steroids, able to churn out text based on minimal human prompts. One iteration, called GPT-3, was trained on 45 terabytes of text data or roughly the entire text available on the World Wide Web. But GPT-3 has no sense of understanding or the meaning of the words it writes. GPT-3 has been used to generate poetry in the style of Emily Dickinson or Walt Whitman or prose like Ernest Hemmingway. It can sometimes be startlingly good and sometimes startlingly bad. OpenAI released its related AI chat interface, ChatGPT, to the public in November 2022, generating a lot of media attention and debate on the uses of machine learning in the workplace. ChatGPT is similar to GPT-3 but much smaller, specifically designed as a chat interface, meaning that it's better suited to understanding questions and commands in chatbot format. The system itself sparked ethical concerns, like the risk of students using it to cheat on schoolwork, and some amazement, as users generated realistic-looking work reports, songs, and short stories. Microsoft then made headlines in January 2023 when it invested $10 billion in OpenAI, presumably foreseeing some future in the technology, as well as profitability.[2] OpenAI is working on updates. GPT-4 is much larger than

GPT-3, with more parameters and an even broader training dataset. Further upgrades to all such systems are highly likely as both the hardware and software improve.

Use in the World

NLP is used in translation, preventing spam, retrieving information (like a Google Search), summarizing text, correcting spelling, answering questions, and analyzing sentiment. The real-world applications of this are chatbots used in customer service, to understand what users are saying or asking, as well as spellchecking for word processing, and search engines for understanding natural language queries. NLG is also used by chatbots and customer service automation to write replies to questions and voice assistants to create answers in real language. It's used to automatically compile financial reports that use the same data points each time. It's also utilized in creating product descriptions and in aggregating and summarizing news and sports reports. Various industries are starting to use natural language machine learning techniques to write reports and analyze data. French bank BNP Paribas launched an NLG service in 2020 to help its custody clients improve monitoring and oversight. The service writes a one-page summary that alerts clients to anomalies and suggests actions, for example, to point out corporate action instructions received after deadlines. Chase Bank improved its marketing effectiveness by running ads through a machine learning tool from AI company Persado. This used an algorithm based on a dataset of ads from other industries and competitors and suggested small tweaks to both the language and look of the ads that increased engagement, specifically encouraging more paperless signups and boosting mortgage applications.

Media companies are also using machine learning to read and write. The Associated Press hired a company called Automated Insights to help automate its news reports for corporate earnings and sports coverage. Automated Insights used data from Zacks Investment Research to publish multiple financial stories in a few seconds. The technology helped AP raise its earnings coverage to 4,400 stories a quarter, a 15-fold increase.

Bloomberg LP, Dow Jones, and Reuters also generate automated financial news at lightning speeds using machine learning systems. Bloomberg has units dedicated to retrieving information and search and discovery, specifically related to the financial information that feeds its terminals. The company uses machine learning for autocomplete, question comprehension and answering, summarization, and relevance ranking.[3] Thomson Reuters has announced plans for a push into generative AI. In sports, AP used a similar principle to produce automated basketball recaps. AP started automating previews of NCAA Division One men's basketball during the 2018 season using data from Stats Perform to deliver over 5,000 previews for regular-season games. The *LA Times* has experimented with automating news about earthquakes. These technologies work by imbibing structured data and using templates to write sentences around it to create stories from specific data points. Narrative Science, an NLG startup that was bought by Salesforce in 2021, is developing software to write tailored dashboards based on companies' business intelligence data. The idea is that words and stories are more useful than data and metrics to most employees.

Technical Details: How It Works

NLP breaks down sentences and words into fragments so the grammatical structure can be understood in context, broken down, and analyzed. The ultimate aim is to have computers read language and understand in a similar way to humans. NLP takes unstructured language and applies algorithms to extract meaning from the fragments of each sentence. Semantic analysis and syntactic analysis are two methods used in NLP to extract meaning.[4] There are various ways to analyze syntax, the grammatical ordering of words in a sentence such that they make sense. Lemmatization means trying to reduce a word to its root word, with the aim of garnering meaning. Stemming is similar but involves chopping off a word to find its root. Another method is morphological segmentation. This involves splitting words into morphemes, the smallest meaningful units of language, which can't be further divided. Sentence tokenization divides blocks of text into its component sentences. And word tokenization is dividing chunks of text into its component words. Part-of-speech tagging means marking a word as it corresponds to parts of words in speech. This is used for breaking down how words are constructed and can be used by machines to categorize how sentences are built, simplifying some of the problems in understanding language. Parsing means breaking down the structure of some text by analyzing the words based on the underlying grammar. Semantics are more complex than syntax for machines because they deal with meaning. Techniques here include Named Entity Recognition — figuring out what the name of the thing is — and word sense disambiguation — gleaning the meaning of a word from the context. NLG is the process of generating words from data. The method for this consists of analyzing content, understanding data, structuring documents, aggregating sentences, structuring grammar, and presenting language. Other related techniques used for text include Markov chains, Recurrent Neural Networks (RNNs), Transformers GPTs as used by OpenAI, and Bidirectional Encoder Representations from Transformers (BERTs).

Importance of Language Systems

Because humans ultimately want to communicate with machines in our language rather than theirs, these natural language systems will become ever more important in our lives. You can already see the trends from the increasing availability of autocompletes, the surge in the use of chatbots, and the rise in no-code development, which lets non-computer scientists build apps or programs. With the Internet of Things, people will want to command their appliances without confusion or ambiguity, and improving NLP is key to this. As natural language computing becomes more embedded in technology, industry, and society, it will be necessary to understand how these systems work and where they are being deployed. There will be a point where people won't realize they're talking to a chatbot or that a robot wrote a poem. One use of NLP is sentiment analysis. This is where text analytics goes beyond extracting words and sentences, to detect positive and negative sentences from text. One use case might be online reviews, helping companies figure out how their products are perceived, and another in finance to parse through company earnings or analyst reports to see how sentiment looks. The outsized public interest in OpenAI and ChatGPT is likely because of the gigantic size of the model on which it's based, and the size of its training dataset improves its performance to such a surprising degree, and the simplicity of its user interface, similar to a Google Search box.

Law firms are also making use of NLP to help with discovery, automation, legal research, document management, contract and litigation analytics, and predictive analytics.[5] The amount of data available to law firms is growing, though it remains largely untapped.[6] Elsewhere across industries, use of artificial intelligence for analyzing natural language is spreading, from fund managers, to insurers, to health care providers, to retailers. And the same underlying technology also powers foreign language translation, which is to where we turn next.

Chapter 8

Lost in Transliteration

In this chapter, we examine how computers translate languages and how machine learning allowed a quantum leap in automatic translation. On a more philosophical level, to what extent can a computer recognize speech? An academic breakthrough in 2016 allowed Google Translate to revolutionize machine translation. In explaining translation, we describe neural networks, and why they are important, and delve into how speech recognition works.

Geoffrey Hinton changed the game for machine translation when he went to work for Google and helped introduce his theories of neural networks. It was one of a number of times he's changed the game for machine learning in his career. Hinton had battled through a prolonged period when the computer science community frowned upon neural networks. In 2012, Hinton was hired by Google as an intern (for purely technical reasons). He went on to lay the foundation that would revolutionize translation software in the real world. Machine translation means having computers automatically translate from one language to another. It was one of the first applications of computers in the 1950s, but the computing power needed was beyond what was available. During the AI Winters of the 1970s and 1980s, the computer science community scorned neural nets because they were imprecise by design. Computer scientists favored symbolic logic over neural networks.

University of Toronto computer science professor Hinton (the great-great-grandson of George Boole, the inventor of symbolic logic) is now regarded as one of the main architects of deep learning, along with his compatriots Yann LeCun and Yoshua Bengio. He spent a decade plugging away at his research when it was out of vogue. In 1986, he worked on a key algorithm called backpropagation, used to test for errors in multi-layer neural networks.[1] In 2006, he made a breakthrough with a paper, "A fast learning algorithm for deep belief nets."[2] He also coined the term 'Deep Learning' that year to describe a specific subset of machine learning. Hinton's practical breakthrough came in 2012 when he discovered one clear use in the real world for his theoretical work: computer vision. But language is different from vision. AI Winters occur when funding dries up after a period of hype and overpromises by the industry. Money freezes, jobs are lost, and excitement in the field evaporates. The first began in 1973, and the second in 1984. Each was precipitated by a differing series of events. The prospect of computers translating between Russian and English instantly was highly appealing during the Cold War. The US government, the military, and academia were excited by the ideas of machine translation. But 1966 saw a series of failures in the field.

The Automatic Language Processing Advisory Committee, or ALPAC, was formed in 1964 to keep track of advances in the field of linguistics in computing.[3] The seven scientists on the committee issued a report in 1966 that was devastating to machine translation. The paper investigated government translation needs, questioned machines' efficacy, and concluded that machine translation was weak and research should rather be directed toward speedier human translation. Funding dried up for two decades. Symbolic logic was seen as the way that translation should work. Language is a system of rules, and logic should be able to dictate those rules, leading to accurate and clear translation. The trouble was that most language isn't logical. It is fluid, ambiguous, and mercurial. So for years, machine translations were quirky, muddled, or downright bad. This was especially so with dissimilar languages, such as translating between English and Chinese. Hinton's approach changed all that. The advantage of deep learning systems, which use neural networks, was that they could better handle the ambiguity of language.

Google's crack unit, Google Brain,[4] was investigating the use of neural networks that learn via trial and error, similar to a toddler, and as such gain flexibility.

Brain developers, led by Jeff Dean, then teamed up with Google Translate engineers to put this new technology to test on the company's whole translation product. The result turned out to be leaps and bounds ahead of what was possible before. And it put the theory into practice. *The New York Times* explained Google's process in detail in an article at the time.[5] The main example of this translation success came comparing two paragraphs from Ernest Hemingway's *The Snows of Kilimanjaro*, translated from Japanese, one using the new system and one the old. They were translated into Japanese by University of Tokyo professor Jun Rekimoto, who then used Google Translate for the English. Results are clear as follows:

Kilimanjaro is a mountain of 19,710 feet covered with snow and is said to be the highest mountain in Africa. The summit of the west is called 'Ngaje Ngai' in Masai, the house of God. Near the top of the west there is a dry and frozen dead body of leopard. No one has ever explained what leopard wanted at that altitude.

Here's the old version, self-evidently a worse translation, which is as follows:

Kilimanjaro is 19,710 feet of the mountain covered with snow, and it is said that the highest mountain in Africa. Top of the west, 'Ngaje Ngai' in the Masai language, has been referred to as the house of God. The top close to the west, there is a dry, frozen carcass of a leopard. Whether the leopard had what the demand at that altitude, there is no that nobody explained.[6]

The release of the new system,[7] which it dubbed the Google Neural Machine Translation system, or GNMT, covered Chinese, English, French, German, Japanese, Korean, Portuguese, Spanish, and Turkish, with other languages supported by Google added later. The initial migration to the new system took nine months. Google released a whole paper explaining the science behind the system.[8] The thrust of why neural networks and deep learning work better for translation than traditional logic or rules-based systems is that they learn from millions of examples. And the system gets better over time. Also, it can intuit the broader context of the sentence, rather than memorizing translations of set phrases. Scaling up the system to 103 languages still represented a "significant challenge" for Google,[9] which needed to figure out how to allow one-shot translation between languages the machine hadn't encountered before.

It wasn't just Google figuring out the use of neural networks for translation. Chinese search engine Baidu was looking into similar methods for its own translation. Andrew Ng, who co-founded Google Brain, later went to Baidu for a spell as chief scientist. Baidu released a pocket translator in 2018, which can be used for simultaneous translation or help interpreters in real time. That device used deep learning and neural networks.[10] Meanwhile, Nvidia was breaking records in training language models.[11] Nvidia is making progress on building larger processors, the hardware used by other technology companies in large machine learning systems.[12] For the future, Google and Baidu have made advances in the models used in translations since these initial breakthroughs in 2015 and 2016. But the quality of automatic translation is still imperfect, as Google scientists themselves admit. There are errors in subject domains, conflating of dialects within languages, and overly literal translations. Informal language and slang are also a problem. But because the systems self-improve and the datasets are enormous, they are getting better all the time. Also, Google is working on near real-time translation, letting its products work as a virtual interpreter. The amazing uses of this technology continue to transform communication. Google Translate has 100 languages[13] and translates more than 143 billion words a day. The most commonly used translations are between English and Spanish, Arabic, Russian, Portuguese, and Indonesian.[14]

Speech Recognition

Speech recognition lets computers process speech into words. It focuses on translating voice to text to allow the machine to identify spoken words and print in machine-readable form. The microphone captures the sound, and the machine analyzes the audio, breaks it into pieces, digitizes it, and then matches it to text. Speech recognition typically includes four features: language weighting, speaker labeling, acoustics training, and profanity filtering.[15] Language weighting lets the machine assign grades to frequently spoken words to improve precision. Speaker labeling tags specific people to identify them in conversation with more than one speaker. Acoustics training lets the computer deal with ambient noise or other background sounds. Profanity filtering identifies keywords to remove cursing or bad language.[16] All these are hard problems to solve. Some examples of speech recognition algorithms are Hidden Markov Models, N-grams, neural networks, and Speaker Diarization. Natural Language Processing is also used in speech recognition. Speech recognition combines with natural language processing technologies in all sorts of useful ways. It can power virtual assistants, it can allow vocal user interfaces, and it can link with translation systems to provide language interpretation in real time, which Google is also working on.

After Luis Van Ahn sold his Recaptcha web technology to Google, he started a language-learning company called Duolingo. The aim was to create a fun language app that draws on the personalization trends that are growing across education. The idea was to make it more like a game than a textbook[17] but also increase retention, a problem for language learning systems. The system uses machine learning to personalize the learning material, serving it at exactly the right level for the student. Duolingo uses a deep learning system via Amazon Web Services.[18] As computers got more advanced and as machine learning improved, robots started to be able to break all the new Captcha puzzles. Google was ready. Recaptcha v3 released in 2018 uses a scoring mechanism to analyze user suspiciousness. Google reckons it can tell humans from robots based on their behavior on the site. The new version also dispensed with the need to identify images in pictures, which humans keep losing, infuriating users and hindering acceptance to sites. Google says it can detect bots far better than the earlier systems with minimal disruption.[19] Critics counter that this system is secure and easy but at the cost of privacy. It's a better user experience, but it gives Google even more control of the web. However, machine learning applications are expanding beyond just language and are getting incorporated into both industrial machinery and household appliances as the technology in sensors improves. One key theme is warehouses, which is described next.

Chapter 9

Movement of the Moment

In this chapter, we discuss huge warehouses, zippy square robots, and managing stock by computer. The upheaval caused by the move to online shopping has elevated the importance of warehouses, and supply chain problems in recent years have compounded this issue. We highlight how algorithms are changing inventory management and how Amazon is revolutionizing product storage and retrieval because of e-commerce and demand for next-day delivery. We also look at item detection and why it's so difficult for robots to sort things into piles and boxes.

A t Amazon's 1.25 million-square-foot fulfillment center warehouse in Schertz, Texas,[1] orange box-like square robots zip around laden with a stack of goods. A giant yellow arm called Robo-Stow lifts pallets up onto shelves with smooth hydraulic precision. But it's not just the robots that are key to driving Amazon's ability to deliver packages to your door within two days. It's the prediction algorithm combined with the bar code technology that lets the company manage its inventory and know precisely where each item is stored to maximize space and speed delivery to customers. Warehouses and packing and distribution systems are now part of modern life as more and more people shop online, particularly after two years of COVID-19 lockdowns pushed consumer behavior toward digitization and doing everything from home.

Warehouses have ramped up using machine learning to optimize their operations, handle staff scheduling, and manage inventory. And artificial intelligence usage at factories, distribution, and supply chain companies is growing faster than other technologies. It's projected to accelerate to 73% within five years, up from 15% adoption, according to the 2022 annual report from MHI, the largest US logistics and supply chain association.[2] As machine learning becomes enmeshed in logistics technology, it's helping with everything from storage, to warehousing, to packing, to inventory. That's allowing the goods to show up at our door within a few days, or even hours, of a mouse click. It's hard for robots to pick out and pack items in the same way humans can. But learning software is still driving warehouse technology. It just means an amalgamation of human and machine, working together in ways that make the best use of both. Machine learning finds patterns in supply chains, identifying flaws quicker and better than humans.[3] It can track inventory levels, quality, and supply and also help in optimizing space. Shippers are getting involved too. FedEx is exploring computer-assisted vehicles, robots, and drones for its deliveries.[4] UPS uses machine learning in an application that lets engineers view global activity and route shipments to the best places according to capacity.[5] DHL deployed an AI-powered sorting robot and is using AI in its logistics operations.[6] Meantime, supermarkets are seeking to better track inventories. Walmart is using machine learning to suggest replacement of out-of-stock items, a problem for consumers and retailers alike.[7] The retail giant scaled back using facial recognition to catch shoplifters in its stores in 2015 amid privacy concerns.[8] Now it's using computer vision to detect missed scanned items in an effort to avoid theft.[9]

Chubby Cuboids

Amazon had around 1,100 distribution centers in the US as of 2022, with 305 of the largest warehouses averaging 800,000 square feet.[10] The robots are called Drives and were built by Kiva Systems, a company that Amazon bought and renamed Amazon Robotics. These squat robots carry piles of shelves directly to employees, who can pluck the items they need, rather than roaming aisles and aisles to search for them. The robots weave a dance, avoiding one another and speeding up the time for the workers to grab the goods they need to fulfill the constant orders. The machines are short, squarish cuboids.[11] The first generation were 1 foot tall, but Amazon squashed them down to 9 inches or less. The next generation of robots, named Hercules, or H-Drive, can carry 1,250 pounds, 500 pounds more than their predecessor.[12] QR codes dot the floors, and the machines have camera readers in their bellies that let them detect where they are positioned in relation to their merchandise. They are controlled by a centralized computer via a WiFi network and use infrared to detect obstacles. When they need a charge, they head to the docking station. The robots are given names rather than serial numbers to help human workers see them as aids and companions rather than competitors. More shelves can be packed in a tighter space, while empty shelves can be brought to packers for restocking rather than standing empty. Amazon uses machine learning in its computer vision systems to detect the objects after they are unpacked, figuring out into which bin they go. Computer models are also used to calculate the size of the box needed for each order and to minimize the distance that the pods need to travel. For the last mile of package distribution to the customer, Amazon also uses an algorithm to optimize how fast items can be delivered. Flex is Amazon's delivery service that uses self-employed drivers to deliver the last mile the goods need to travel.[13] The last mile of delivery has long been a thorny problem for the industry. In terms of machine learning, the trick is in combining machine speed and scale and human dexterity to accurately retrieve and pack customer orders. There are one to four million product bins per fulfillment center, and Amazon is constantly iterating to assess in real time which orders should be picked and when, to ensure the products that go together in the same box. The company uses computer vision systems to tell where each item is located in the warehouse and track its position, according to Feedvisor.[14] The warehouses are set up in a Manhattan-style grid system, with structured paths for the pods to follow. Each pod typically has nine rows of shelves on each of four sides. The warehouses range in size from 600,000 to 1 million square feet (more than 15 football fields). Amazon sees fully automated warehouses as at least a decade away, and robots are working with humans at 26 of the 175 fulfillment centers. But the combination of humans and machines is increasing.[15]

SLAM

Amazon has also installed machines called SLAM, which stands for Scan, Label and Manifest. The SLAM system weighs the parcels and compares them with an expected weight, with any discrepancies checked by hand. During packing, machine-readable barcodes are attached to each parcel, linking it to related information. But SLAM robots scan the barcode and, using air pressure, add a printed label that can be read by a person. SLAM can handle its tasks without stopping the parcels, saving time. Amazon warehouses use a 7-step process: receive, stow, pick, flow, pack, SLAM, and ship. The items are all below a certain size, allowing them to be sorted upon retrieval. Amazon also has a separate system for large items, such as kayaks or large TVs, which are handled by no-sort warehouses, or what it calls non-sortable fulfillment centers.

Warehouse Wonder

Amazon isn't the only company using machine learning at warehouses. Alibaba, the giant Chinese sourcing supplier, has a logistics arm called Cainiao that is using Internet of Things technologies in its warehouses to automate and streamline operations. Ocado, an online UK grocery supplier, says it has the most automated warehouse in the world. It uses six disruptive technologies for its platform and has over 200 patents.[16] AI, robotics, digital twins, cloud, big data, and Internet of Things are combined to enact this vision. The combination lets customers place a 50-item order in minutes, from specific standpoints, such as organic or gluten-free shopping.[17] It uses machine learning to predict how quickly items might run out. The company also uses the technology in its own demand forecasting to optimize for freshness, minimize waste, and handle inventory. There are more than 58,000 items available for sale on the Ocado platform. Still, Ocado is having an identity crisis as to whether it's a retailer or a software company, according to analysts, and needs to make its mind up to determine future strategy.[18] It certainly acts as both, though in one sign it's moving toward software; US supermarket giant Kroger adopted Ocado's platform at its Kentucky site in 2022.[19]

Robot Workers

The way roles will shift in warehouses and logistics centers depends on the direction the technology develops, and understanding the workings of the new systems to be adopted will be key, specifically overseeing new platforms and troubleshooting. Robotics will also be key. Employees will need to both predict these trends and bridge the gap toward automation by garnering the skills to oversee, manage, troubleshoot, and maintain the systems getting put in place in warehouses. In the future, warehouses will rely further on various digitization technologies, but humans will still be required for hybrid tasks and supervision. Fulfillment centers will get more efficient and improve logistics and tracking processes. Leveraging artificial intelligence will help with managing inventory, timing, and space.[20] While warehouses have been affected by the rise of e-commerce, the COVID-19 pandemic put pressure on supply chains and threw up holes and weaknesses in some of the industry's practices.[21] The Internet of Things applies to warehouses as well as the home, and smart connected devices will be key. The pandemic revealed cracks in the supply chain infrastructure. And technology will be used to repair and improve that. Next, we turn to the home, where the use of automation has taken off in one specific area: robot vacuum cleaners.

Chapter 10

Roomba Rhumba

In this chapter, we explore gardening robots, paper robots, machines with soft hands, and what the Internet of Things will do for the home. As technology lets more and more devices link wirelessly to each other and to central controllers, people are opting for ever more convenient gadgets. We study why the robot vacuum cleaner is indicative of things to come. Devices embedded with sensors and connected to each other allow increasing digital control around the home. You can operate your blinds, thermostat, television, and security system all from one place. Does the home of the future look like Tony Stark's house?

Roomba co-creator Helen Greiner was inspired to build robots by Star Wars droid R2–D2.[1] There's evident similarity between the autonomous vacuum cleaner and the world's most famous fictional droid in looks. The Roomba is a circular beeping robot that sweeps the floor, with the advanced models calculating room size, assessing a sweeping path, and using sensors to stop itself from falling downstairs. They are among the most successful robots ever, with parent company iRobot selling more than 40 million robots in total as of 2021. (In fact, Amazon agreed to buy the company for $1.7 billion in 2022.)[2] How is it that the first robots to be widely available for home use turned out to be vacuum cleaners? Why not something more aspirational or at least robots with broader and more personal applications than just sweeping? When considering the path of the technology, this development actually makes sense and ties in with the often narrow uses of machine learning. It's much harder to build multi-use learning models, just as it's much harder to build multi-use robots. And it also emphasizes the difficulties with making Artificial General Intelligence as against single-use systems.

Roombas can sweep a room unaided, navigate obstacles, and return to base for a charge. Competitors to iRobot vacuums are also available. Roborock, Eufy, and Neato all make similar offerings with differing technology. They all solve a consumer problem in that few people enjoy vacuuming. Autonomous vacuums can be used far more often than is practical for their human-aided counterparts; they're quieter and save considerable time on a mundane household chore. Pets can ride them; they're just a cool device to have around. In general, robots struggle with emulating human dexterity and precision. In warehouses, even with the advances in technology and learning, tasks are increasingly stratified according to the limits of the tasks of which robots are capable.

Paper Robots

For example, strawberry picking is notoriously hard for robots because of the soft nature of the fruit and the need for non-mushed strawberries. But robots are increasingly being used in agriculture. In Amazon warehouses, machine learning and automation are on the rise, but as we've seen, humans still have defined and crucial roles and will continue to for years to come. This ties in with the view of many academics that machine learning and automation must be designed to work in collaboration with humans, rather than to simply replace them or take over their roles. Instead of replacing humans through mechanization, roboticists should figure out how to get people and robots to work together, says Daniela Rus, head of the MIT Computer Science and Artificial Intelligence Lab (CSAIL). Machines are better at crunching numbers or lifting heavy objects, but humans can reason, generalize, and work in the abstract. Working in combination, they can augment each other.[3] Roboticists are increasingly rethinking what robots should look like and what they should be made of. Rus, for example, has worked on ingestible origami robots made from paper[4] and soft robotic hands that can safely grasp delicate objects.[5]

The key for workers is in figuring out how automation and technology will transform their jobs, along with the roles and requirements, and then adapt their skills to fit those changes. In the home, wired and smart products are advancing apace. The category of robot vacuum cleaners fits into the broader technology of Internet of Things, which is rapidly altering how people use devices around their home. Lighting, smart appliances, blinds, thermostats, and locks can all be controlled centrally through apps and mobile phones. Machine learning technologies are part of what makes this possible.

Best Robot Ever

But back to vacuum cleaners. The Roomba is probably the most commercially successful robot ever, certainly the most popular house robot. The most advanced Roombas are now able to map their surroundings, using sensors to orient themselves. They are also able to navigate pitfalls, such as stairs. They also have memory banks to remember the location of furniture or other obstacles. They learn as they go along, storing a virtual map of your home.[6] iRobot was founded in 1990 when Greiner and fellow student Colin Angle teamed up with their professor, roboticist Rodney Brooks. Angle went on to be CEO of iRobot and Brooks later founded Robust.AI, where he built a new kind of cart-like warehouse robot called Carter, designed to work in close proximity to people, and is another example of rethinking the robot paradigm and stereotype.[7] Carter looks more like a cart than a robot.

Lawn Care Robots

Greiner went on to design the PackBot, used in bomb disposal, exploration, and to search through debris in disaster situations, like after 9/11 or in the aftermath of the Fukushima earthquake and tsunami. She then became CEO of gardening robot company Tertill. Often described as Roomba for the garden, Tertill is a weeding robot, whose special wheels churn up the earth to stop weed seeds growing. Robot lawnmowers are another, related, area of commercial success in robotics.[8] The market is growing. Swedish chainsaw maker Husqvarna built the world's first robot lawn mower as far back as 1995. But robot mowers have increased in popularity as people increasingly look to automate gardening tasks, the COVID-19 lockdown spurred more interest in lawn care, as well as innovation in the industry driven by wired-home technology. These robotic mowers can be controlled remotely and sensors stop their blades automatically if needed in emergencies. Companies are trying to harness the latest technologies such as GPS mapping, machine learning, and solar power.[9] iRobot has its own lawnmower, Terra. Back to vacuums again, the Roomba uses an optical sensor to snag more than 230,400 data points a second, letting the robot create an accurate map of your home, "so it knows where it is, where it's been, and where it needs to clean."[10] Roomba uses cameras. Some competitors use laser navigation, or Lidar.

Floor Level Pictures

iRobot upgraded its systems in 2020 to give the Roomba an overhaul. Called iRobot Genius Home Intelligence, it is able to target a specific area for cleaning. It is also able to detect obstacles, like cables, or even soft pet waste, which are all tricky hazards for autonomous vacuums. The way the Roomba works is by using computer vision to take pictures and create a map of homes. iRobot needed to train computer models using pictures from floor level because such images are so different from photos taken elsewhere. iRobot collected tens of thousands of images inside employees' homes to help learn what furniture looks like, for example. The company even used pictures of play dough to emulate pet poop, an interesting dataset to be sure.

As homes become more wired, machine learning will play an ever bigger part in home appliances and become ever more integrated into systems. Products such as Nest and Alexa can hook up speakers, computer games, WiFi, smoke alarms, door locks, and cameras. Nest explicitly looked into using smart alerts to detect motion or activity on camera and send an alert to your phone.[11] This trend is only going to accelerate, leading to demand for smart items around the home. Amazon is likely to increase compatibility between Alexa and Roomba with the acquisition linking up many more wired home devices. Alexa uses a technique called "active learning" to improve its responses. This method cuts down on the manpower needed to train its models by focusing on the most important bits of missing information. Alexa uses models trained on data annotated by people. The data are labeled to help the machine classify the questions people ask Alexa. For example, if you say "Play the Prince song 1999" or "Play River by Joni Mitchell," the labels of ArtistTitle and ArtistName are tagged to the song and the artist. This allows the machine to classify similar queries without needing annotations. Amazon focused its efforts on training the examples that will improve Alexa the most:

> But annotation is expensive, so we would like to annotate only the most informative training examples — the ones that will yield the greatest reduction in Alexa's error rate. Selecting those examples automatically is known as active learning.[12]

As is clear, this technique uses natural language understanding to interpolate questions in English. More and more tasks and applications will need natural language machine learning techniques as technology requires simpler interfaces, which let people communicate with machines with verbal questions or commands. As the models improve, the applications will get better, and the interfaces will get cleaner.

Robot Hinderers

One observation is how strikingly good virtual assistants are at some questions and how strikingly bad they are at others. This ties in with the tasks computers are good at and the tasks they tend to struggle with.

One advantage of machine learning systems is that they improve over time and with more data. There's tension between academia and industry, especially as it ties into the Internet of Things, because commercial needs and practicalities are so different from the theories of what is possible in the realm of the smart home. From this disconnect surfaces opportunities in both spheres, harnessing the best research that can have practical applications. Miniaturization is also likely to be a theme around the home and the warehouse. For now, it's hard to fit huge neural networks on tiny devices.[13]

As far as jobs go, the Internet of Things is opening up a whole new field. As devices move online, jobs will be required on both the hardware and software sides. Data scientists specializing in data from home devices will become more in demand. New devices will be created, as well as old devices with new interfaces, such as smart fridges or dishwashers. Not everything is panning out like futurists predicted. The idea of a generalized smart home helper robot is getting more remote. Artificial intelligence is moving toward narrow specific tasks rather than broad generalness. The idea of a home helper like Iron Man's Jarvis is still in the realm of fiction. But applications like ChatGPT, Siri, and Alexa are increasingly going to be able to control our devices. The future of the smart home is clear, with more connectedness of more smart electronics, leading us to our next topic and one that is crucial to machine learning: sensors.

Chapter 11

Sensory Overload

The world has recently experienced rapid advancement in sensor technology which, when combined with machine learning, opens up powerful possibilities in technologies from medical diagnostics to self-driving cars. In this chapter, we analyze why this combination of technologies is such a watershed and outline where sensors are used, why they are proliferating, and how algorithms and models are driving their adoption. A particular area of interest examined here is the way in which sensors are combined with artificial intelligence in the miniaturization of machine learning, not only for our smartphones and watches but also for areas as diverse as gesture recognition, contamination detection, and factory production lines.

S ensors are used in medical diagnostics, power sources, factories, and self-driving cars. So as the wired world around us further integrates, they are inextricably linked with machine learning. What exactly are sensors, why are they needed, and what do they do? Sensors detect a physical property, such as light or sound, imbibe it, and provide a reading that corresponds to changes in the world. They are used in abundance in cars, phones, and household appliances, at greater or lesser degrees of sophistication. Inputs can be sensed in the form of sounds, vibrations, electrical signals, light, or temperature. Sensors are needed to relay information from the devices we use to capture and measure data.

And sensors are proliferating as the world becomes more connected. Meanwhile, hardware is shrinking and sensors are getting cheaper. Consider all the sensors in your smartphone. Many smartphones have motion sensors. Accelerometers measure linear acceleration, while gyroscopes detect rotation speed. These, for example, allow the phone to react to being shaken. Apple phones use this as an undo command, as well as to act as a pedometer and track how many steps you've taken today. What do sensors look like? Among the most visible examples are motion sensors used in home security. Those linked to burglar alarms are perched in the corner of a ceiling, often with a square white eye housed in a white plastic casing. Most sensors just look like part of electrical circuits, according to their function. So how are sensors connected to machine learning? It turns out machine learning is becoming integral to sensor design. There's also a growing distinction between tiny self-contained sensor technology and huge machine-learning language systems.[1] Machine learning in sensors is becoming known as tinyML. This is where sensors are a self-contained unit with an embedded machine-learning system trained on a specific task, such as recognizing gestures or listening to audio.[2] They can be run on much smaller, cheaper micro-processors with tiny batteries, at a much lower cost than larger systems.[3]

Tracking Speed

Sensors are also everywhere in cars. They measure speed, fuel level, oxygen, airflow, and fuel temperature and watch for proximity to objects to aid parking. Self-driving cars use sensors even more, to help them "see" the road and the world around them. Autonomous vehicles combine cameras, radar, and Lidar to construct three-dimensional images and sense their environment. They also have sensors called inertial measurement units to track speed, acceleration, and the distance of objects. Cameras are placed on all sides of the cars, giving them 360-degree viewpoints. Camera-based sensors struggle in low visibility, night-time, or fog. Radar, most commonly known for how it tracks aircraft or ships, sends radio waves in pulses. Radar helps supplement cameras in cars.

Lidar takes it a stage further by measuring distance using pulsing lasers. It provides shape and depth perception. Where machine learning comes in, is that driving is incredibly difficult to program in a rules-based way. There are so many exceptions, surprises, and common-sense actions taken for safety, that to program a car to drive is an almost limitless open-ended task, as we will see in our chapter on self-driving cars. Machine learning lets the car learn from experience and data without being explicitly programmed for specific tasks. This is crucial. In self-driving cars, sensors are still unreliable in bad weather. So long-distance trucking is simpler than driving in New York, for example, but harder than driving in a retirement community, where speed is slow. Roboticist Daniela Rus uses the example of driving through a mountain pass in the Rockies during torrential rain, whereupon you'd need a strong collaboration between the sensors and the control system.[4]

As the Internet of Things becomes more prevalent and embedded in our lives, sensors become more and more key. They are also needed for wearable devices, and the linkage between the device and the data storage and analysis mechanism consists of sensors. Sensors are also used in healthcare, lighting, and factories. Sensor data are typically sent to what is known as an edge computing device. Edge computing allows devices at the edge of a network to monitor and store data, providing faster insights that are closer to the data. Sensors are crucial to almost all aspects of modern life, including safety, monitoring, process control, as well as diagnostics and public health, according to the Electrochemical Society. In the future, sensors also will be able to self-monitor, self-correct, and self-repair. Developments are in the works for sensors to see with photonics, feel with physical measurements, smell via electronic noses, and hear using ultrasonics. Also, in the future, sensors will combine with actuators for movement and with smart electronics and wireless for communication.[5]

Toxins and Fluids

Smart sensors have built-in computers letting them conduct predefined tasks, before processing information and passing it on.[6] Such smart sensors are fundamental to factories under Industry 4.0, as the Fourth Industrial Revolution has become known. They allow factories to work autonomously by monitoring equipment and measuring temperature, levels, sound, and others. Gas sensors monitor air quality, chemical sensors detect toxins, and level sensors measure quantities of fluids.[7] As both devices and models miniaturize and incorporate systems, there's a big upheaval taking place in embedded sensors. As one academic paper puts it, "Machine learning sensors represent a paradigm shift for the future of embedded machine learning applications."[8] On the negative side, embedded machine learning sensors are complex to integrate, are not modular, and suffer from concerns about privacy. These are all concerns that the industry must address as it innovates and builds out the technology at an ever smaller scale.

Whither Sensors?

As of 2023, 7.33 billion people globally have mobile phones.[9] The sensors in these billions of cell phones are going to become more and more important as data-first attitudes take hold and as it becomes easier to work with sensors. As the wired world integrates, the Internet of Things grows more mainstream, and cars move toward driving themselves and sensors will become embedded in our lives. The learning technologies which go along with that, and which are crucial to interpreting the data, will be needed, along with user interfaces to allow users to interpret and analyze this sensory information. When you attach a fitness app to your phone, you can analyze your health data in real time. When you drive a car, the Lidar, radar, and cameras are translated into maps or video feeds, or other apps on the dashboard. Smart sensors are cropping up in various places. One group of Canadian researchers built a sensor system that detects bacterial contamination in water based on a visual image. It could classify *E. coli* in water with 99% accuracy.[10]

With the advent of smartwatches and health trackers, people are carrying around a plethora of sensors without even realizing it: cameras, infrared sensors, microphones, optical sensors, ultrasound sensors, accelerometers, and gravity sensors.[11] These data can be analyzed via machine learning for such diverse tasks as recognizing human activity, fuzzy classification, detecting failure, calibrating cameras, controlling telescopes, identifying objects, assessing wildlife, and auditing shelves. With the world so awash with data, the needs grow to house the data, as well as wrangle and interpret it for everyday use. Machine-learning techniques are crucial to do this, as the data are often unstructured and in different formats and from different platforms and devices. Of course, privacy risks grow with every piece of data stored, especially regarding medical or health information. But this also throws up yet more opportunities in terms of new companies to handle this kind of thing, along with new jobs to tackle the task.

Engineering Sensor Jobs

Sensor technology is another area of the workforce where skills are lacking. There are many new types of jobs being created as the technology around sensors grows more advanced, specifically in hardware such as phones or cars and in software. Sensors bridge engineering and computer science and intersect hardware and software. As with elsewhere in technology, there's a growing demand for mechanical engineers who can design and build the requisite infrastructure to handle sensor systems and a shortage of people to fit those roles. And also like elsewhere in tech, cybersecurity experts are in demand.

In future, sensor accuracy will increase, and sensors will become smaller and smarter.[12] That in turn will improve safety, and measurements will get better. Sensors will power more and more of our devices around the home and office. The growing demand for measurement in modern society boosts demand for sensors. There are sensors for soil, for climate, and for capturing real-time data using global positioning system (GPS) tracking technology. Wearable devices, health tech, and data-driven approaches all require sensor measurement, and all are generating more and more data. Machine learning can quickly interpret and analyze such data and provide users with instant insight.[13] Next, we move from the world of sensors to the world of the workplace, where machine learning is about to have far-reaching consequences.

Chapter 12

Nation of Workplace Automation

Automation in the workplace is a common fear among employees. In this chapter, we explore how ever cleverer machines are being used to automate tasks and processes. We focus on what robots and machines can do, and what they can't, and how employees must adapt as roles and requirements change. Machine learning is not synonymous with cutting-edge workplace automation, but there is likely to be an overlap. Finally, we also consider what a Universal Basic Income is and whether it will be needed.

Automation theorists offer a utopian reply to our dystopian world.[1]

— Aaron Benanav

The great fear of the age is that robots will take your job. But will they? Fears about workplace automation and consequent loss of jobs have been around for decades. But now, companies across industries are implementing automated systems at scale, building machine-learning departments, and figuring out how their many areas of clerical work can be handled more efficiently by machines, with fewer staff and no lost productivity. That's because companies are cutting costs generally and looking to shed workers. In this chapter, we'll examine how this transformation is taking shape across industries and identify the skills needed for workers to stay cutting edge. It's always better to be controlling the robots rather than being replaced by them.

Where's Your Job Going?

Emeritus states the following:

Workplace automation refers to the use of systems to perform repetitive or predictable tasks without direct human inputs. Automation can be applied to physical tasks using machinery or robots, or to data-driven processes using software and algorithms.[2]

Some 47% of US jobs are at high risk from automation in the coming decades, according to a study from the Oxford Martin School.[3] Much white-collar office work in the 1990s and 2000s consisted of punching numbers into spreadsheets, cutting and pasting text from one place to another, and other generally rote work. The advance in technology means that, in theory, this kind of work can now be automated. In practice, it's not so straightforward, even as executives push for it. Most jobs are not 100% rote work, meaning the practice of clerical automation is complex; some things require thought and strategy, while others are simple and replicable. Automation machines are far more task-oriented than humans. The same is true of factory work, where people can still handle many tasks far easier than robots, even though the push to automate production lines has continued apace. Smart companies realize how to maximize the uses of both humans and machines and play to the strengths of each to maximize productivity. There's precedent in the history of farming. The share of total employment in agriculture dwindled to less than 5% in 1970 from 60% in 1850, according to management consultant McKinsey. New kinds of jobs were created, cushioning the economy from the loss of farm work, even as food production boomed in the period.

Economists and executives have considered the effect of workplace automation for years and the subsequent ethical quandaries. While some physical manufacturing jobs have already been automated, the takeover by robots looks very different from the scenarios that were earlier feared, as can be seen in our chapter on warehouses; workers are still required and will be for some time. But their jobs are changing. Now, a paradigm shift in the automation of clerical work is upon us. The same syndrome will apply, in that the disruption will take a very different form than what those same economists and executives think, unless they really understand the advantages and the limits of the technology and can foresee technological trends.

High Machine Costs

The pros of automation are clear to business owners, executives, and managers: less human error, an increase in scale, and a slashing of costs with automated work and the theory that human workers can be diverted to focus on higher-level and more interesting or useful tasks. Better employee engagement is possible if automation is implemented right; few workers enjoy repetitive tasks.[4] Automation can also increase consistency and predictability. If set up correctly, automation can mean less risk of expensive errors or, in dangerous manual tasks, fewer physical injuries. And the main benefit is scale: Automation systems can scale up across organizations. The cons include potential job cuts, the shifts in worker roles and demands, and the need to upskill and retrain new and existing employees. There's also a certain loss of flexibility from robotic systems. People inherently fear and resist change, especially where automation is concerned, complicating implementation efforts. High costs of development and of putting in place computer-aided automation systems are another con that is often underestimated by executives, both in terms of engineers' time upfront and maintenance once systems are up and running. Further drawbacks include extra implementation costs, such as licensing or vendors. The roles of workers will change, too, meaning reskilling and training. Finally, automation systems can be inflexible or brittle.

One new industry that has sprung up in recent years is called Robotic Process Automation, or RPA. This involves virtual robots who squat on clerical workers' computers, watching to learn their repetitive tasks, and then replicating them and taking over. The idea is that workers can train their personal "robot" to take care of much of the spreadsheet wrangling, data entry, calendar entries, and other replicable tasks. This same technique can be used across industries such as law, healthcare, banking, insurance, and others. UIPath, a leading robotic process automation company, uses the tagline: "We make software robots, so people don't have to be robots."[5] UIPath is aiming to use machine learning to enhance the robot processes while offering prepackaged models that handle tasks like classifying text in emails or web pages, relaying customer queries to frequently asked question documents, understanding sentiment in product reviews, or classifying images. One example that UIPath offers is to automate email processing using a template to remove the drudgery of reading, prioritizing, and acting on emails. Rival companies Automation Anywhere and Blue Prism offer similar systems of desktop robots. This kind of white-collar automation is taking place in marketing (sending messages in response to an action), human resources (handling payroll, expenses, and timesheets), managing inventory (tools to track and predict inventory levels), data entry (scraping data, cleaning data, and automatic uploading), and customer service (chatbots, frequently asked question answerers). Obviously, machine learning and AI systems are needed to implement all these with any usefulness. It remains to be seen how much of a revolution this specific type of automation causes for white-collar work, but many of these functions are already handled by machines and in varying degrees of quality. Who else will be affected in the coming years? Predictable physical work, data processing, and data collection are among those roles most susceptible to automation, according to McKinsey. By contrast, managing others, applying expertise, making decisions, planning, and using creativity were found to be among the jobs least affected by automation.[6] But executives have been surprised before and consultants' predictions don't always come true.

Control Your Learning

The World Economic Forum's (WEF) concept of the Fourth Industrial Revolution, where technologies fuse and lines blur between the physical, digital, and biological, means that "technology-related and non-cognitive soft skills are becoming increasingly more important in tandem."[7] The WEF urges employees to take control of their own lifelong learning and career development to preempt this transformation. Wise advice for sure, as companies ditch policies of hiring lifelong employees and labor economics shift. Various industries are poised to spend more and more on machine learning, and concurrently many sectors will see increasing adoption by businesses. The debate often focuses on evaluating automation that augments jobs versus automation that replaces jobs. The hard fact is that both types of technical changes leave workers without jobs, according to Benanav in *Automation and the Future of Work*.[8] He uses the example of a retailer replacing four checkout cashiers with self-checkout machines, with one employee overseeing the system. Is this the end of cashiering, or is one cashier now operating four machines? This change has occurred rapidly in the retail industry in recent years, as the technology got cheaper and better. Supermarket self-checkouts basically hand over the work to the customer, and while cheaper for stores, don't always improve the customer experience. Usability is often an issue and can be an example of the importance of clean user interfaces.

Skills Mismatch

If worker automation in the workplace continues to the extreme, wiping out the labor force as we know it, some economists say we need a Universal Basic Income (UBI). That's the theory of paying a base level of money to all adults to help people survive in a society without jobs. Computer scientists and economists are divided on the need for this. Proponents argue that it will be necessary as more of our work gets automated by technology.[9] Automation will make so many tasks obsolete that it will remove the need for many jobs. Critics of UBI say that work often defines people's identities and creates meaning and connection, and so replacing it with a government check isn't enough. Also, a basic income is highly expensive, fails to address training needs, and could worsen inequality rather than reduce it.[10] MIT economics professor David Autor points out that while many tasks of "middle-skill" jobs are susceptible to automation, most such jobs demand a mixture of skills, which can't easily be unbundled for machines to take over in full. He cites the growth in medical support roles, such as radiologists and nurse technicians, as an example.[11] The more conceptually complex and diverse your job, the safer you are from automation, or being replaced by machine. Also, workers who can quickly learn to operate the machines poised to replace them are a clear advantage, as they can sidestep upward to higher-function jobs. This holds for both manual jobs and white-collar work. Wouldn't you rather be in control of the machines? Chatbots, applicant tracking, space optimization, and text generation for writing reports are all ways automation is moving into the workplace. There's a shift from customer relationship management to customer data platforms. There's also a skills mismatch for many of the roles in the modern workplace. Companies complain they can't find the talent they need in certain specific fields, even as they seek to cut costs and shed staff.

Phone Tree Horror

Customers are likely to push back where automation doesn't adequately fulfill the roles played by human employees. Think of the process of endless automated phone trees while trying to reach a live customer service agent at many large corporations. Also, some questions are nuanced and simply can't be answered by machines. Consulting firm Deloitte conducted a survey in 2020 to garner a global view of how organizations implement and scale intelligent automation technology. They polled executives from countries across the world and in a range of industries. It turns out that the speed of innovation is outpacing the speed of adoption. Deloitte pointed to these main obstacles: fragmented processes, lack of IT readiness, lack of clear vision, and resistance to change. These are clear problems for companies, but opportunities exist for employees looking to take advantage of the shift to automated systems. Clerical automation is taking place in law, healthcare, finance, and insurance. More physical automation is occurring in manufacturing, agriculture, and food service.

MIT professor Autor uses the paradoxical example of bank tellers, who have doubled in number in the 45 years since the introduction of the Automated Teller Machine (ATM) from about a quarter million to half a million. Why hasn't automation eliminated their employment? While ATMs replaced tellers, they also made it cheaper for banks to open branches, leading to more tellers, but handling less routine work. Most work we do requires a multiplicity of skills, and in general automating some subset of those tasks doesn't make the other ones unnecessary, it makes them more important, Autor says.[12] While not every role is rendered obsolete by machines, there are a plethora of new jobs that will be required as companies move toward data-driven strategies and digitized operations. Meanwhile, executives can get carried away and underestimate the skills of humans while overestimating what automation is good for. Elon Musk famously said humans are underrated when describing excessive automation at Tesla.

New jobs are being created across multiple industries in data analysis, data science, data management, and data engineering. Also key are the implementation of digital strategies, building and managing systems, and leading data and engineering teams. The key for workers is to acquire the future skills companies need for the ongoing digitization and the coming data revolution. It's in companies' interests to help their employees skill up and craft roles that optimize how machines and humans can work together, or they will be short of the tasks and roles they need to stay cutting edge and struggle to attract and retain the talent they require. One area that faces among the biggest disruptions from machine learning and artificial intelligence in the years to come is healthcare, to which we turn next.

Chapter 13

Image Problem

In this chapter, we analyze how machine learning is used in diagnostics and medicine, such as for predicting heart failure and lung cancer. We identify some of the problems found in such a complicated effort and the likely outlook for artificial intelligence in healthcare. While there's optimism regarding the future of medicine, it's still much harder than it looks to mesh this technology with the untidiness of the real world, especially in such a complex and fragmented industry, a lesson from which executives in other sectors could learn.

Can a computer diagnose lung cancer early? One of the most promising applications of machine learning is in the healthcare industry. There's so much scope for usefulness, and it could be a real game-changer. It's also one of the areas of the biggest disappointment in terms of the realities and the difficulties of using technology in the real world. Healthcare data are hard to work with, and the complexities of how the industry operates snarl its use. As we'll see, there is cause for hope as scientists and clinicians increasingly start to understand the capabilities and limitations involved.

Diagnostics from medical imaging is one such area. The idea is that with a large and accurate enough dataset, machine learning models can become better at predicting diseases than doctors, for example, detecting lung cancer from X-rays. One reason is that machines are much better than people at spotting patterns from large datasets and can even find instances that doctors might miss, given enough historical data. As with elsewhere in other sectors, the quality of the data is crucial. The further upside is that doctors could use such systems to enhance their own diagnostic techniques. The technology really has the capacity to revolutionize parts of diagnostic medicine. Still, some high-profile efforts have fallen short. Here again, a constantly recurring theme in artificial intelligence occurs where the hype doesn't match the reality. And, similar to self-driving cars, the stakes are higher in healthcare than in other industries as people's health and safety are affected or lives are at stake. This chapter will look at the ways machine learning is being used in diagnostic medicine, particularly through images, how it has the capacity to revolutionize the industry, and why certain efforts have failed. Then we'll look at the future prospects based on the current state of the art.

Diseases and Drug Discovery

The main uses of machine learning in medicine are for the diagnosis and identification of diseases, drug discovery, and medical imaging diagnosis. It's also being used in clinical trial research, crowdsourced data collection, and to personalize medicine. The main obstacles and hurdles within healthcare are data governance: Medical data are private and governed by regulations. Also, the transparency of algorithms remains an issue that hampers adoption. If your algorithm is a black box, it is likely to meet resistance. In drug discovery, drug development regulations require transparency. Also, electronic records are fragmented and unstructured. Without central structured databases, machine learning is almost impossible to implement. Medical data exist in silos and drug companies are resistant to changing this. Finally, there aren't enough data scientists with domain expertise in the healthcare industry to match demand.

IBM Health: A Case Study

IBM had high hopes for using its supercomputer, Watson. The idea was that Watson could revolutionize healthcare, but IBM's Watson Health division failed to live up to expectations after IBM arguably focused more on marketing than on the implementation and faced resistance by practitioners. IBM eventually sold the unit for parts in January 2022, after a decade of underperformance.[1] When IBM's famous Watson supercomputer appeared on the game show *Jeopardy* and beat the human champions, IBM used the event and the global publicity around it to tout Watson and its computing power as having a real-world application, specifically in healthcare. In 2015, IBM's then-CEO Ginni Rometty declared that healthcare was IBM's next big thing. "Our moonshot will be the impact we will have on healthcare," Rometty said at the time. It turned out to be harder than she thought.[2]

IBM Health, which used Watson to generate insights from medical data for doctors, was a classic example of AI hype overtaking reality. IBM overpromised and the Watson-based system under-delivered. IBM Health at one point had 7,000 employees, and the company invested billions in buying giant health datasets on which to focus the technology. Hopes were high when IBM partnered with Memorial Sloan Kettering Cancer Center in New York in 2012 for oncology research. The aim of the project was to offer doctors better access to cancer data and provide individualized diagnostic and treatment recommendations for patients.[3] At another IBM partner, the University of North Carolina School of Medicine, the academics and IBM workers were at odds over the capabilities of the technology and what it could do. "We thought it would be easy, but it turned out to be really, really hard," Dr. Norman Sharpless, former head of the school's cancer center, who went on to be the director of the National Cancer Institute, told *The New York Times*.[4] "We talked past each other for about a year." At the time of the sale, IBM said it remains committed to Watson and is focusing on cloud and AI strategy. The healthcare data and analytics were sold to a private equity firm.[5]

The problems with IBM Health are common to misunderstandings of AI usage elsewhere, namely interpolating a narrow usage to an unrealistic wide scope and also underestimating the breadth of a problem and ascribing too much optimism to models that are incapable of handling such complexity. Winning *Jeopardy* was an extremely narrow and tailored use that was completely different from health uses.

Despite the setbacks in healthcare, there have been some successes and learning. Algorithms are helping doctors detect cancer, heart disease, and Alzheimer's, and AI improves the hospital experience and helps prepare patients to go home to continue rehabilitation faster.[6]

An Eye on DeepMind

All this is not to say healthcare is a lost cause for artificial intelligence. It's just necessary to understand the limitations and the best uses of learning models and algorithms. DeepMind entered a partnership with Moorfields Eye Hospital in London in 2016 related to eye care.[7] The project was slow at first. It took two years for the first phase, eventually showing the algorithm could match doctors to recommend treatment for 50 diseases. Doctors had used optical coherence tomography, or OCT, scans to look for irregularities. It took the doctors time to examine these complex eye scans, and the machine could cut down on that time by using two neural networks to identify 10 features of eye disease.[8] The next stage was helping clinicians predict eye diseases before the symptoms set in. The system interpreted eye scans and suggested how patients should be treated, potentially cutting the time between scan and treatment, helping those with age-related macular degeneration and diabetic eye disease.[9] Researchers trained the model on datasets of anonymized retinal scans from 2,795 patients with wet AMD, a serious version of the disease that can make patients go blind.[10] Moorfields went on in 2021 to develop an AI system to detect geographic atrophy, a chronic degenerative condition affecting 5 million people globally.[11] The upshot will potentially expand the techniques across 30 hospitals in the UK, and Moorfields says the projects will help wider research for years to come in the country's National Health Service. Globally, more than 285 million people live with some form of sight loss of which eye disease is a major cause. Such disease can be prevented with early detection and treatment. It can be used for finding cardiovascular anomalies, detecting fractures, diagnosing neurological conditions and thoracic complications, and screening common cancers.[12] Here's another example of how machine learning can work with medical images: Brain diseases are some of the most costly, in terms of suffering and economically.[13] Convolutional Neural Networks or Recurrent Neural Networks could be used for neurological images in a more quantitative approach to analyzing neurodegenerative disease to improve decisions. It can enhance screening and improve precision in medical images generally.

Hearts and Lungs

Heart disease kills about 17 million people globally a year. Lung cancer is the leading cause of cancer death in the US, the main cause of cancer death worldwide for men and second-most common for women. Both these diseases show promise from using machine learning predictions. Most lung cancers are only detected at a late stage.[14] If people can be screened earlier and high-risk patients identified before symptoms show up, they can be treated faster and their longer-term outlooks vastly improved. This is where machine-learning systems come in. Learning models also can be used to rule out patients with benign lung nodules, making the process of screening cleaner.[15]

Only about 7% of those with metastasized lung cancer are still alive after five years, according to Johnson & Johnson.[16] But it's difficult to diagnose. Symptoms can be vague, such as coughing and fatigue. The types of data used in prediction systems can in future go beyond imaging data to include genomic data, medical records, and even cellphone and wearable data, according to researchers in a paper examining lung cancer. The value of image analysis won't completely replace biopsies but will give extra information that is complementary to them. Also, the models need to be more transparent to show clinicians why decisions were made. However, they reduce human error and cut costs. Using AI systems will create a "new era in cancer management."[17] Also, the problems with integrating these tools into workflows of radiologists means that automation is likely to be adopted in incremental steps over time. That means clinicians won't have to radically change their methods of checking for lung cancer overnight. And computer-assisted decision-making already exists for lung cancer screening, meaning that new tools will slot in gradually.[18]

No One Best Model

Several machine learning algorithms are in use to detect heart problems, although no consensus has yet been reached on the best type of model to use for cardiovascular disease.[19] One meta-study showed that Support Vector Machines, or SVMs, are better at finding hidden patterns in complex clinical datasets. The researchers found that SVM and boosting algorithms have been used to the best effect in heart medicine.[20] However, further studies comparing conventional risk models with machine learning analysis are needed. There are also some systematic hurdles in medicine slowing down progress, including limitations of the data, bias, and research incentives. Also, growth in machine learning research doesn't automatically lead to clinical progress; in fact, some reports show it has seldom had clinical impact so far.[21] In one example cited, of 62 published studies on machine learning for COVID-19, none had potential clinical use.

The future of using such models in diagnostics is hopeful, even after all these challenges. As the technology improves, it will be able to learn with smaller datasets and training data and with fewer annotations, a key issue when accounting for the precious time of doctors and surgeons. The lack of medical data science skills presents an opportunity for industry experts to learn data interrogation techniques and for data scientists to learn about medicine.

What's Next in Medicine?

Medicine and healthcare are the business categories that received the most private funding in AI in the past five years, garnering $28.9 billion, taking it ahead of data management, financial technology, and retail.[22] That suggests investors are betting on some of the knottier problems in the sector getting solved with machine learning. Many practitioners believe healthcare is the place where the technology will have the biggest societal impact. Farther into the future, futurist Ray Kurzweil predicts nanobots will get injected into the bloodstream to boost our immune system.[23] There are more pressing issues to solve for now in medicine, as well as some ethical and privacy considerations to contend with, which is the topic of the following chapter.

Chapter 14

Ethics of Data Lakes

In the following, we discuss what the explosion in data means in terms of ethics. Machine-learning models are now being used in the real world in a concrete way, and problems are arising quickly. Once algorithms are up and running, it's too late to implement moral frameworks, so people need to think about big data's impact on society now, not later. How do all these technologies tie together, and what are the ethical concerns that are involved? Such questions span privacy, the military, and exactly who is responsible for the underlying algorithms.

Ethics is getting big in big data. The sheer amount of data in the world and the pace that it's expanding are raising ethical concerns across the board. How personal data are used, how algorithms are prone to bias, and the lack of transparency of machine learning systems all raise moral questions for the technology industry and for broader society. That's in addition to worries about superintelligence, autonomous weapons, and facial recognition. Satellites track us, phones geo-locate us, cameras watch us, and credit cards map our financial path through the world. Point-of-sale credit card data, mobile phone texts, information from car sensors, and social media posts are all contributing to the barrage of data available for machine learning models to crunch through. The ethical implications of this are real. The amount of data in the world is expected to surge to 175 zettabytes in 2025, from 33 zettabytes in 2018, according to IDC estimates. That's a stack of blue-ray disks (if you remember them) that could reach the moon 23 times.[1] What that means is that the world is awash with data. This has an impact on you in terms of privacy and in terms of how algorithms based on such data are having more and more impact on people's day-to-day decisions. As more tasks become automated, decisions are outsourced to learning systems that are at the risk of bias because of their underlying data or because of the way their models were set up. Also, you might not know that you are interacting with an artificial intelligence system — not a risk with the current state of customer service chatbots, but watch this space. Machine learning is the tool that allows the harnessing of this ocean of data: the sifting, the filtering, the sorting, the tracking, and the predicting.

How Transparent?

Ethics in data science is a relatively new phenomenon. As the systems have grown more powerful, and the amount of data available in the world increases, the risk of abuse and bias rises. Both the technology industry and the general population are only starting to worry about these issues. And governments are slower still, but they are starting to pay serious attention. Such ethics are inextricably linked to both education in computer science and the design of algorithms by technologists. Without taking into account ethical considerations, models are open to abuse, will contain bias, and may behave in unexpected and unpleasant ways. With this astounding growth of data in the world comes risks. Where is all these data coming from, and how is it getting created so fast? Hedge fund analyst-turned-data scientist Cathy O'Neil points out the risks to such a data-driven society in what she terms Weapons of Math Destruction, or WMDs.[2] To O'Neil, big data must never be unthinkingly relied upon and can even be a threat to democracy if such bad algorithms are allowed to proliferate unchecked.

Privacy has already become a hot topic globally, but with the advent of facial recognition and location detection technologies, it takes the problem to another level. Countries can monitor their citizens, companies can monitor their customers and workers, and big data companies can monitor everyone. Algorithms have taken over many of our tasks. Machines can read our faces.[3] Finally, machine learning or artificial intelligence can be so entrenched in our technology that it's almost invisible, meaning it's harder to track or judge its impact. If people don't know machine learning is embedded in the app they are using, the social media site they're interacting with, or the customer service chatbot helping to answer their question, does it matter? Yes. Does this raise ethical concerns? Yes. Four of the biggest moral worries in big data and machine learning repeated by academics and industry experts are those of privacy, bias, opacity, and security. There's also the need to train large datasets properly.[4] Tied into these are also the tangential worries of larger existential threats, the elimination of jobs, and the development and buildup of autonomous weapons. Opinions are highly divided about which ethical issues should be the key focus. These range from alarmists such as philosopher Nick Bostrom, who worries about the development of superintelligence, to academics-turned-practitioners who argue that job losses due to computers and machines are a bigger risk. Machine learning pioneer Andrew Ng contends unemployment is the industry's biggest ethical concern. Some companies make ethics a central tenet, like OpenAI, which promises Artificial General Intelligence that benefits humanity, and DeepMind, which is explicitly committed to safe artificial intelligence. Part of the deal when DeepMind was bought by Google was to set up an AI ethics board. Microsoft is most worried about facial recognition and killer robots, the two issues that generally raise the concerns across computer science academia and industry, which we have examined in earlier chapters. Machine learning advances play a big part in both of these.

Google's "Don't Be Evil" mantra becomes thornier when you consider the amount of data the company has at its disposal and how much of the world's search it controls. But Google at least has its AI Principles, which aspire to work for the common good. The contrasting attitudes of Google and Microsoft to accepting US military contracts highlight an ethical dilemma. Google dropped out of the US Defense Department's Project

Maven[5] AI surveillance system after more than 3,000 employees penned a letter saying Google shouldn't be in the business of war.[6] Microsoft, which picked up part of the contract, said it believes in a strong US defense, and the military should have access to the best technology available.[7] Based on international relations realism theory, this is also a strong argument (depending on your nationality and political viewpoint). But the point is these questions aren't easy. Other moral quandaries surrounding deep learning and exponential data increases are cropping up daily both for the technology companies that employ the techniques and collect the data, and the governments, which regulate the companies, and must stay up to date on the cutting edge. In one stark example, Huawei, the Chinese telecom company sanctioned by the US for spying, filed patents for facial recognition technology that recognized Uighurs, a Muslim ethnic minority in China, from pictures of pedestrians.[8] Human rights organizations say China discriminates against Uighurs, rounding many into forced labor camps; China says that its camps are voluntary and used for training. It's not just China. Moral questions abound in machine learning more so because of the mystique of systems that teach themselves. People are wary of autonomous machines, and even though academics say some commonplace worries are misplaced, they identify other areas of concern. As usual in industry, regulation is lagging behind innovation. But regulations are surely necessary for such revolutionary technologies to prevent harm or abuse. Even the global technology behemoths say it. Governments say it. And consumers are certainly calling for it. The UN is looking at a framework for intergovernmental agreement on the use of autonomous weapons, as we saw in an earlier chapter. This is one of the most difficult ethical areas of machine learning, where life and death are at stake and machines are able to make decisions without human interference. But facial recognition technology is, if anything, more advanced than lethal weapons.

Keep It Private

Privacy is hard in an era of big data. And different countries have different attitudes toward, and laws affecting, privacy. But the fact that all these data are getting collected all around us means that privacy is a question relevant across the globe. The General Data Protection Regulation, or GDPR, is a law put into effect by the European Union in 2018, which affects companies and organizations worldwide.[9] The rules are designed to protect the data privacy and security of EU citizens and are a clear reaction to the advent of big data. They levy big fines where companies do not comply and greatly increase the administrative burden for how companies handle customer data.

Machine learning lives off data. Without data, and the more granular the better, models will founder and fail. So there's a constant tension between the gathering of datasets and people consenting to their data being harvested. Health data are particularly sensitive, and many data science projects use anonymized datasets for that purpose. Social media companies have increased transparency about the data they collect and how it's used, both as a reaction to GDPR and at the demands of their users. The Cambridge Analytica scandal, where 87 million Facebook profiles were scraped without users' knowledge for psychological profiles used in targeted election campaigning, also made consumers question the usage of their data. Cambridge Analytica's implosion, and the publicity of just how targeted social media mining can be, really brought data privacy into the mainstream.

Bias in the Model

Bias in machine learning has become a big topic as it was realized that there are major ethical considerations by those who make the algorithms, which might not have been understood at the time they were made. Microsoft built an AI chatbot called Tay in 2016. Designed to interact with the public on X (previously, Twitter) or through messages, the system quickly turned racist and sexist as users taunted the model. Tay tweeted 95,000 times within the first 16 hours, with a worrying percentage of those tweets racist or sexist, forcing Microsoft to apologize and switch the chatbot off.[10] The bot was built on neural networks that allowed it to learn from its interactions. Unfortunately for Microsoft, human trolls from the Internet site 4chan bombarded the bot's feed with harmful language, teaching the system to internalize the comments. That meant it spewed hate speech unprompted. More recently, Facebook's AI Blenderbot made anti-Semitic comments and even described its own CEO Mark Zuckerberg as creepy and manipulative. The training data, along with interactions with the public, also probably caused the problem. Meta acknowledged the chatbot may make offensive remarks and said it was still experimenting.

It's not just bias in language systems. Google's photo app in 2015 mistakenly labeled black people as gorillas.[11] Again, the company apologized, but fixing the algorithmic bias wasn't easy. The source data were to blame. The classification system removed the words gorillas, chimps, and monkey from Google photos' captions, which was not a real fix, and AI designers were forced to go back to the drawing board. Biased models have a real impact on society and are something machine learning scientists are aware of and addressing. Sometimes the bias is subtle, sometimes less so. Predictive policing is an example where race can also spark controversies because of the underlying crime data on which the models are based. This technique focuses policing resources on where and at what time crimes are most likely to be committed. In lending policies, if there's a scarcity of data in one section of society, for example, where cash use is higher than credit cards, that would push interest rates higher for those people compared with where credit card use is more widespread. This could have a demographic or racial impact, even if unintended. Who shapes these policies? It's often the algorithm designers. These problems show that ethical considerations should always be considered at the initial stages of building models. At an earlier stage, ethical considerations should be embedded in computer science education. Further, algorithms and machine learning should innately and by definition contain ethical soundness. Many of the ethical quandaries facing machine learning are applicable to broader digitization and technologies. However, transparency and the black box problem are somewhat unique to machine learning. The reason is that as algorithms become more complex, for example, with an increase in the layers of a deep learning or neural network, the decisions to get to a specific output are harder to pin down by a human. This is a problem as explainability is necessary for certain applications. And getting people to adopt the system becomes harder if it's not possible to explain how some outcomes were reached.

Security and Future

Finally, security is always a high-profile issue. What happens if your data are stolen? What happens if a company's database is stolen, affecting millions or hundreds of millions of customers? The systems need to be smart enough to fend off attacks. For safe and useful machine learning, moral reasoning should be an integral part of what it means to do data science. We need to change training systems to account for this. Mark Coeckelbergh argues in his book, *AI Ethics*, that such ethical concerns should be embedded into artificial intelligence education.[12] We need to bridge the gap between arts and science to overcome this "lack of inter-disciplinarity," he says. That means those with a humanities background need training in new technologies, what they do, and where the risks lie. Scientists and engineers should likewise be taught the ethical and societal aspects of software development and its use. Ethics shouldn't be seen as a marginal topic but as ingrained in artificial intelligence, constituting an essential part of it. Many practitioners think of ethics as a constraint, but it could support long-term sustainable development. We must not allow the mindless use of algorithms. Once they are loose in the world, it's too late.

Chapter 15

Economist, Forecast Thyself

In this chapter, we discuss how economists are using machine learning to plug statistical gaps and improve their predictions, but are also worrying about how machines handle causal inference. Many researchers are just starting to see the wider uses of the technology in this field. Artificial intelligence is also changing the economy itself, displacing jobs and altering the workforce. Economists foresee an array of scenarios where machines replace humans in the economy, requiring a societal rethink of what it means to work. As we analyze, some say a Universal Basic Income will be needed to maintain a balance as jobs are displaced or vanish.

Economists, who have long used statistics as their stock in trade, should have been primed to be at the forefront of machine learning. But the economics industry has been slower to adopt the technology than elsewhere, and practitioners have been reluctant to use tools because models typically can't handle causation or answer questions about correlation. That's now changing. One of the key benefits of machine learning to economists is the ability to work with bigger datasets. Statisticians often say machine learning is just a form of statistics anyway and that they have been doing it for decades. What tends to distinguish the two is a background of economics and statistics versus a background of computer science by practitioners. Regardless, machine learning will transform how empirical work is conducted in economics, according to Susan Athey, a professor of the economics of technology and an expert in both.[1] Such changes will include the use of off-the-shelf models for specific tasks, like prediction or classification. There will also be a great deal more interdisciplinary research, and consequently, the way funding works will shift. That's because research on giant datasets requires specialist skills and is labor intensive.

Computer scientist and professor Fei-Fei Li argues for precisely such a multi-disciplinary approach in machine learning, where economists are consulted on jobs, large companies, and finance, with policy makers and legal scholars brought in to talk about regulations and ethics. Li argues that computer science alone can't solve all AI's problems. She seeks a kind of human-centered AI, based on three pillars: AI should be based on human impact, should enhance humanity, and should be inspired by human intelligence.[2] For Athey, with further digitization across various sectors of the economy, policy advice will be needed across farming, education, health, and government services, all of which create the need for economists to propose and build out frameworks to help the various sectors. Societal effects of machine learning will also require study and research from economists. Machine learning can be used for economic forecasting predictions and modeling and to help minimize error. This can augment economists' existing methods. It can also handle far larger datasets and multiple models at once. Alternative data are another route that machine learning can be harnessed to enhance economic forecasts. Geospatial data, credit card data, shipping data, and other datasets that are much larger than government economic indicators can help economists predict more traditional statistics, often in a more timely manner. Companies including Safegraph, Quandl, Kensho, and many others have built new businesses just by providing alternative data in recent years. There's also been consolidation in the sector as large data providers snapped up alternative data startups. (US stock exchange Nasdaq bought Quandl and data company S&P Global acquired Kensho.)

There's also an economic technique called Nowcasting, used to predict gross domestic product (GDP) in real time. This uses higher-frequency data than traditional economic statistics and indicators.

Nowcasting can incorporate alternative datasets and leverage machine learning to improve the accuracy of the predictions. The Reserve Bank of New Zealand studied machine learning for predicting GDP in a 2018 paper, which concluded that real-time Nowcasting with learning models outperformed statistical benchmarks and could further be improved by combining models. These new techniques represent a useful addition to the toolkit of economic prediction, the study found. Along similar lines, Google's DeepMind has used real-time Nowcasting techniques to forecast the imminent chance of rain over the next one to two hours.[3] This is one of the National Severe Storms Laboratory's Grand Scientific Challenges and is a big enhancement to traditional forecasting in meteorology. In more traditional economic usage, the Atlanta Federal Reserve, along with some other regional central banks, has developed Nowcast models to predict US GDP. The Atlanta Fed describes using monthly source data, with a method called Bayesian vector autoregression, aggregating the 13 subcomponents that make up GDP, and using similar methodology to the Bureau of Economic Affairs.[4] These Nowcasts are now closely watched by market participants because they provide a speedier update than the government's official GDP figures.

Economists also examine leading indicators to get a head start on how certain economic statistics will pan out. Analyses of these kinds of data, as well as newer alternative datasets, have long been the preserve of hedge funds and investment firms, as we see in our chapter on asset management. Macroeconomic data can help in asset management decisions. Economists are also starting to use the same kind of techniques but with an eye to better explain economic linkages. Sentiment analysis is another area where economists are using the new technologies of natural language processing as a more subtle gauge to extract meaning from text. Parsing of central bank statements, for instance, where the language changes subtly over time, and where machines can ingest and analyze hundreds or thousands of words at high speed, means the machines can spot patterns that humans can't see. Market participants use it to gauge the sentiment of central bankers and policy makers, extracting instant meaning from speeches, releases, and minutes of meetings.[5] Banks also use machine learning to hone forecasts, economic predictions, and modeling and to minimize errors. Central banks themselves even use text mining to examine the broader economy, and interest in big data has surged at central banks in the past five years. They also use it for regulation and supervision.[6] Machine learning offers new tools that can provide a framework that helps economists better understand growth and business cycles, says James Bang in a book exploring just that subject. In particular, it helps gauge the likelihood of recessions.[7] Economics professor Athey foresees a number of big modifications in how economists will conduct research:

> My prediction is that there will be substantial changes in how empirical work is conducted; indeed, it is already happening, and so this prediction already can be made with a high degree of certainty.[8]

Universal Basic Income

Then there's the underlying economy itself and the ways that machine learning and artificial intelligence are drastically changing the shape of jobs across the spectrum and are likely to cause further upheaval in the future. If, as this book contends, machine learning is a transformative technology throughout society, then there's a strong need for economists to examine this too, especially if economic and workplace changes are as sweeping as some predict. Economists have started to express interest in this technological watershed, along with the concurrent changes to the labor market. Computer scientists and machine learning practitioners are already worried about the economic impact and fallout from their creation.

Economists such as Erik Brynjolfsson have written about the effect of technology on the economy and specifically the effect of artificial intelligence on productivity. Brynjolfsson argues that automation and augmentation can boost productivity and wealth, but the growing pie won't necessarily be shared. He argues that we should work on challenges that are easy for machines and hard for humans, rather than hard for machines and easy for humans.[9] This is an excellent observation. This trap of automating for the sake of it is a risk to executives and can burden companies with high software costs on dubious projects where machines do a worse job of replacing human tasks. Automation, if nothing else, needs to be thoughtful and fit for purpose. As automation in the workplace increases, spurred by machine learning and other related kinds of technology designed to replace human rote work, economists are also mulling an idea called Universal Basic Income, or UBI, as we saw in the earlier chapter on workplace automation. As well as the practical matter of workforce displacement, there's also a philosophical question about how we fill our days with no work to do. The point at which robots or machines take over the bulk of paid work in society brings up the question of how people fill their time, as well as how they get paid, and what they do all day. Some economists argue that abundance will be reached due to technology eroding the marginal costs for such services and is a rosy view of the economy and society. In UBI, citizens are paid a minimum wage for living and won't need to participate in paid work unless they want to. There are also more metaphysical questions that arise about self-worth and meaning if and when technology removes the need for the majority of jobs in the workforce. There's the darker view of course, where this age of abundance is never reached. Some analysts are worried about automation taking jobs away without replacing them, and other economists are less sanguine about the prospects for the leftover workers, either before such abundance is reached, or when it fails to materialize. In this case, politicians and economists need to examine the impact on the labor market and make some decisions, or risk social unrest. Meantime, data scientists are more and more in demand for jobs, as are big-data specialists. This demand is growing fast. According to Addepto's *Machine Learning in Economics*, "as Indeed.com shows we can observe a 29% increase in demand for data scientists year over year and an almost unimaginable 344% increase since 2013."[10]

There's disagreement over how these broader economic changes take shape and what form they will take. Machine learning is already intertwined in the economy via our technology and increasingly in our financial infrastructure. And finance is one area where machine learning is at once new and old, as the following chapter shows.

Chapter 16

Lords of Finance

In this chapter, we examine how hedge funds have used machine learning techniques for many years, staying ahead of the pack in finance. Now, the rest of the financial industry is catching up by analyzing alternative datasets, improving high-speed trading and risk management, interpreting documents, and creating content. What's next for the Lords of Finance? And if this technology is so smart, why can't machines predict the stock market?

Famously secretive, and founded by mathematics genius and former code-breaker Jim Simons, Renaissance Technologies is on many measures the most profitable hedge fund of all time. It's also one of the earliest and most successful examples of hedge funds using machine learning and artificial intelligence in finance. So if machine learning is so smart at predicting, why can't it forecast the stock market? It turns out financial markets are examples of chaotic systems that are too large and with too many permutations for accurate forecasts. That hasn't stopped the finance industry from trying, hedge funds in particular. Simons' genius was that he didn't try to predict the market direction. He used algorithms to sniff out and quickly profit from market wrinkles.[1] Renaissance, formed in 1982, gathered together a group of mathematicians to use statistical methods, examine market pricing, and detect anomalies. It started with commodities, currencies, and bond markets and then later moved into equities. Renaissance typically avoided Wall Street types for its staff, in favor of scientific geniuses and experts in other fields who could bring their knowledge to finance.

So hedge funds, like Renaissance and rival DE Shaw & Co., have used machine learning techniques for many years. Now the rest of the financial industry is catching up, in analyzing alternative datasets, improving high-speed trading, risk management, and document interpretation, and creating content. Machine learning in finance is growing at a rapid clip as financial firms and asset managers seek to harness the latest technology to stay cutting edge, save costs, and adapt to changing customer needs. There's also traders' perpetual hunt for the "edge" that helps them beat the market. As with other sectors, the increase in data and computing power has made machine learning in finance suddenly more appealing. Top uses include tweaking processes to get more efficient, enhancing and strengthening portfolios, limiting risk, and underwriting lending. Adding to the pressure, financial technology and blockchain firms are seeking to disrupt the whole financial industry, providing a growing existential risk to incumbents, as upstarts have the ability to build their entire technology from scratch using the latest techniques. In this chapter, we will focus on investing and asset management, and the following chapter examines banking. Hedge funds used statistical modeling and quantitative data analysis in their trading over many years but exactly how tended to be a closely guarded secret.

Alternative Data

Alternative data are economic signals that offer faster and broader insights into both stock markets and economies. They range from satellite data to mobile phone sensor data, to data from credit card transactions, and because such information is huge and unstructured, machine learning is crucial to analyzing it. Alternative data are distinct from the usual economic indicators or earnings reports that Wall Street analysts and investors have relied upon for years and can be used to supplement such financial information "fundamentals" and hopefully give an early glimpse of the performance of companies or economic performance. Examples include satellite data showing the activity of factories in China, parking at malls, point-of-sale data from credit cards that measure retail spending, and mobile phone data that can show the location of shoppers. This is all newly minted information and often suited to machine learning because of its size and unstructured nature. It can supplement the quarterly earnings reports and economic indicators that are the bread and butter of the finance industry. Alternative data can also be timelier than traditional ways of judging companies and the economy, giving users a head start. The problem is interpreting it and understanding what it's showing. Unlike corporate sales or profit figures, often alternative datasets aren't clear-cut about how companies are performing and need some interpolation.

US investment bank JPMorgan says that investors are keen to know how machine learning is being used in finance and how the techniques differ from traditional quantitative investing: "Significant growth in the use of machine learning to trade financial markets is making ML (Machine Learning) difficult for investors to ignore — especially those looking for new sources of uncorrelated alpha."[2]

Uncorrelated alpha means the extra financial returns from assets over a given benchmark but which is unlinked to other market performance, and so it spreads out the risk of such a trade. Edge, in other words. There is still a considerable degree of work needed to wrangle and clean alternative data, and more to deliver useful or actionable insights from any particular dataset. This has turned out to be harder than expected for many use cases. Financial datasets are often noisier than in other realms. Some hedge funds focus just on statistical analysis of asset price data. JPMorgan says that quantitative investing funds started to adopt machine learning when they realized it could help find predictive relationships and signals, monitor changes to see when a source of return fades, and adapt to change on an ongoing basis. While it can deepen and broaden the investment process, relying on machine learning also contains some risks, for example, overfitting, crowding, and leverage, as well as exogenous shocks, also known as black swan events.

Back at Renaissance, Simons wanted to systematize trading because he didn't like the emotional rollercoaster involved with losing money. He eventually succeeded. Renaissance's flagship Medallion fund made an estimated annualized return of 66% before fees in the three decades from 1988 to 2018. "The real thing was to gather a tremendous amount of data," Simons says.[3] "We had to get it by hand in the early days. We went down to the Federal Reserve and copied interest-rate histories. In a certain sense what we did was machine learning. You look at a lot of data and you try to simulate a lot of predictive schemes until you get better and better at it."

Renaissance's Medallion focused on short-term strategies. Simons and his crew built models to find statistically significant anomalies in markets. One interesting difference with other funds and financial outfits is that Renaissance didn't care why the anomalies occurred. That is, they were unconcerned with causality or the economic reasoning behind the moves, as long as they traded to make money. Another difference is that they built one big trading model instead of separate models for each asset class. That allowed them to suck in data from all kinds of sources to test, long before their competitors in finance started to use such techniques. Finally, they built industrial-scale production-level systems to handle the trading. Renaissance uncovered possible predictions related to market volatility, how two securities relate to each other, known as pairs trading, and other anomalies or inefficiencies. The goal was to parse historic pricing to find sequences that could repeat and then take advantage of that until the opportunity was gone. Medallion fund became "something of a data sponge," soaking up any and all information that could be quantified and scrutinized. Gregory Zuckerman explains the background and details of exactly how Simons went about all this in his book, *The Man Who Solved the Market*.[4] Their wide use of information to trade across different assets enabled the managers at the fund to benefit from far more data than rivals and other Wall Street institutions, years before most people even became aware of big data or machine learning.

Satellites show geolocations, data from sensors, text from conference calls, bills of lading from shipping, heavy truck sales, website scraping, and more; all these measures let the funds infer and interpolate the performance of companies and of the broader economy before the release of official government statistics or corporate earnings reports. This lets funds bet on stocks, bonds, and currencies in different countries and markets with greater confidence. So quantitative hedge fund firms, with this experience, stand to benefit in this new world of alternative data, although as such strategies become known, it's harder to profit, and trades become crowded. One problem in finding profitable trading signals is that they vanish once other financial market participants find them and pounce on them. That's one reason for the secretiveness of hedge funds about their strategies and insights.

Applying machine learning effectively to financial markets is extremely difficult. Alternative data are unwieldy, and market pricing is hard to analyze. That's as fundamental company data and economic indicators have become more transparent and available to all, making trading edge harder to find. JPMorgan's report on hedge funds found that money manager selection and portfolio construction are key to implementing such strategies successfully. That is to say, you can't just rely on the numbers. That's echoed by Stefan Nagel in his book studying machine learning in asset pricing.[5] While techniques in machine learning are well suited to prediction, and the data explosion and better technologies are aligning, Nagel points out that it's still key to tailor models specifically to finance. He concludes that learning methods are useful to analyze asset-price data as well as underlying drivers of demand. They can also be used to gauge the expectations of investors and macroeconomic forecasters. This isn't straightforward. "The idea, sometimes associated with ML, that one could make predictions in an entirely data-driven automated fashion is too good to be true," Nagel says.[6]

That seems to be exactly what Simons and Renaissance did over several decades, although not without some major obstacles along the way. Not many others in finance have succeeded. Nagel also argues that off-the-shelf applications of machine learning tools are unlikely to work well in financial asset pricing without "injecting some degree of economic reasoning." Asset markets are constantly undergoing structural changes. Most machine learning models aren't designed for that and are too brittle to handle such upheaval.

DE Shaw, where Jeff Bezos worked before he founded Amazon, is another fund that used machine learning from the early days, while Two Sigma is a data-science-focused hedge fund founded by a computer scientist and a mathematician that says it aims to bring science to finance. Both have been successful in using machine learning in their investment strategies. Man AHL, based in London, is one of the world's longest-running and largest systematic hedge funds. Man has been using machine learning in its multi-strategy client fund since 2014. It says such systems combine a wide variety of weak information to turn them into stronger financial signals. But there are challenges to data modeling and noisy financial information.[7] Goldman Sachs Asset Management uses machine learning to adapt to changing data in valuing companies. It also uses Natural Language Processing to gauge investor sentiment in news or research reports.[8] Applying these techniques to finance is by no means easy, according to JPMorgan's report on asset management. That's especially true if funds expect percentage returns above single digits. Elsewhere in finance, companies are putting artificial intelligence to simpler tasks, like customer service and marketing. For banking, it's a matter of return on investment, while staying at the forefront of technology and striving to hire the right staff away from technology companies. Finding software engineers with financial experience — or traders with coding experience — is a challenge both for asset managers and banks. Hedge funds (or successful ones at any rate) don't seem to struggle to the same degree in hiring or retaining such specialist workers. But banking's biggest use of machine learning techniques is in combating fraud and money laundering, which is our next topic.

Chapter 17

Finding Fraud

In the following pages, we analyze the various new ways in which the banking industry is using artificial intelligence and why, how machine learning is harnessed in fraud detection and loan applications, and why it's so good at this particular function. Other cutting-edge use cases in banking include customer service applications, risk assessment, and improved clerical productivity. We go on to discuss how jobs in finance are changing to fit the new wired world.

How do you sniff out money launderers? One way is to have machines scan masses of data and flag up anomalies. Criminals launder $2 trillion to almost $4 trillion a year globally, according to World Bank and IMF figures. Illegal cash proceeds of terrorists, drug traffickers, corrupt politicians, and mobsters are all fed into the legitimate financial system by launderers, and banks are on the front line in fighting it, under growing pressure from governments and intra-government agencies.[1] Financial firms are turning to new technologies, and among them is machine learning. US banking giant Capital One says that traditional methods are outdated, so adopting new techniques is essential. Maintaining the growing number of rules needed in a rules-based detection system is hard to maintain. Capital One began deploying machine learning to monitor suspicious activity in August 2020 to adapt to criminals faster, improve governance, and give better insights to investigators.[2] It paired data scientists with anti-money laundering experts.

For banks, the main applications of machine learning are in fraud detection and prevention, compliance, risk management, investment predictions, customer service, and loan underwriting. Banks historically were slower than hedge funds in deploying machine learning. They are now rushing to catch up and future-proof their businesses. They have also struggled to scale the technology from prototypes and experimentation to global or firm-wide use, often a tough task because of their size. According to consultant McKinsey, the main reasons for this failure are inflexible technology investment, lack of clear AI strategies, fragmented data, and outdated operations that prevent collaboration across teams.[3] The management consultant says that for some banks, adopting AI quickly and across the business will no longer be a choice but a "strategic imperative." Regardless, banks are now racing to employ computer programmers and data scientists with a keen eye for implementing the latest machine learning techniques. The trick is to find technical staff that also have knowledge of financial markets. This crossover of domain expertise with machine learning software skills is a prized skillset. Banks are keen to cut costs and eliminate some of the grunt work of the back office, as in trade processing and settlement, and also of the laborious work of loan assessment. McKinsey argues that banks must become AI-first companies, as the technologies can deliver $1 trillion of value per year to the global banking industry. Lenders haven't embraced this kind of attitude in the same way as technology companies. But banking is complicated and regulation-heavy with the added burden of keeping client money safe. The technology ethos pushed by Mark Zuckerberg of "Move Fast and Break Things" is typically a bad strategy in finance.

Sniffing Out Fraud

Machine learning is really good at sniffing out fraud. With millions upon millions of transactions, routine business operations, and money moving in and out of accounts and across borders, fraud detection would seem to be the ideal use case for data science. Transactions are too numerous to police by hand and fraudulent transactions are hard to spot. Machines can sift through the mounds of data at comparatively little cost, and models can be trained to pick up on anomalies, which can then be flagged up for assessment to check on fraudulent transactions. The Bank of England says anti-money laundering and fraud detection are well-established use cases because "the need to connect large data sets and undertake pattern detection is a set-up that lends itself well to machine learning."[4] Financial institutions predict machine learning will be the biggest help to their businesses in the two areas of fraud and anti-money-laundering, the Bank of England found in a survey in 2019. Those two were followed by gains in operational efficiency and finding new analytical insights. Financial firms are increasingly adopting machine learning techniques in the United Kingdom, and of the 300 firms surveyed, many had passed the initial development phase. Banks formerly used rules-based systems for sniffing out fraud.[5] Transactions of unusual size or location were typically assessed according to certain criteria and flagged up. These systems were labor-intensive and time-consuming. Also, they are less subtle in the patterns detected. Some banks are now combining learning systems with rules-based techniques to get the best of both worlds. For anti-money laundering, banks are able to analyze entire customer bases using unsupervised machine learning techniques and then identify outliers for more precise identification of risks. This lets them scale the systems with much less manpower.[6]

Banks have also added new analytics as extra filters to existing systems to cut down false positives, combined traditional approaches with graph theory, and used supervised machine learning to conduct the monitoring itself. (Graph theory is the branch of mathematics that studies the properties of points and lines, widely used in search engines.) Some lenders are experimenting with a "possible paradigm shift" to a holistic approach that monitors customer behavior, according to a report from the Institute of International Finance.[7] The Bank of England said one large unidentified British bank was able to reduce the amount of false positives within its money laundering checks by as much as 70%. This is key as eliminating false positives cuts down on wasted time.[8] Challenges for machine learning in catching money launderers include the quality of the data, hurdles in sharing the data, legacy or old information systems, and uncertainty regarding support from regulators. The Bank of England itself made AI a priority topic in 2022, focusing on the impact of financial firms deploying machine learning on the financial system.[9] The central bank said there may be new risks from the technology along with the efficiencies gained.

Erica America

It's not just about fighting crime. For banks themselves, customer service and chatbots have been a particular focus. Bank of America rolled out a pilot of its AI virtual assistant, Erica, in 2017.[10] This chatbot works by learning from conversations with customers and uses what the bank describes as advanced analytics and cognitive messaging. The bot, which works as a mobile app, has access to your cash flow, balance, transaction history, and upcoming bills and is able to leverage the data to answer customers' questions and help them stay on top of their finances. It can alert clients that cash balances are nearing zero, point out changes in credit score, lock or unlock cards, get account information, and manage payments. Use of the service climbed to more than 1 billion interactions as of October 2022.[11] JPMorgan and Wells Fargo also have trialed chatbots. Wells Fargo tested a chatbot via Facebook Messenger in 2017.[12] JPMorgan launched an automated virtual assistant in its treasury services division for corporate treasurers in 2018. That was aimed at a different segment of the bank's clients, and while the concept is similar to retail products, the usage is very different and shows the different ways banks are thinking about deploying such interactive technology.[13] JPMorgan is actually testing out a number of bots.[14] Another system, COIN, short for Contract Intelligence and used to interpret commercial loan agreements, reduced the amount of work from 360,000 man hours down to a few seconds and with fewer errors, according to the bank. But despite the hype, bots and chatbots are only a small part of machine learning in finance. The biggest use in the financial industry is still for fraud detection and anti-money laundering.

Credit card processor MasterCard rolled out a fraud prevention system based on artificial intelligence in 2016 called Decision Intelligence. The service reduced instances of credit card customers being declined for legitimate purchases. Again, false positives for fraud are a huge time-wasting problem for credit card issuers as well as banks. The system examined user behavior over time to predict abnormal spending patterns. MasterCard engaged an AI firm called Brighterion in 2006, later acquiring the company in 2017 and folding it into its artificial intelligence efforts. Now, 2,000 companies use Brighterion software, and 74 of the 100 biggest US banks use its AI systems. Rival credit card processor Visa has a system called VisaNet +AI aimed at helping customers with authorization, clearing, and settlement. The system is designed to address pain points such as delays, confusion regarding account balances, and the unpredictability of daily operations.[15]

Future of Finance

Banks' other usage of learning technologies includes detecting loan application scams, identifying theft, and helping with loan underwriting. The way machine learning is used in each application differs, but the underlying models are broadly similar. Another reason that Moving Fast and Breaking Things doesn't work in banking is security. The public rightly expects banks to keep our money and data secure, and their business rests on that premise. That means that when adopting any new technology, they need to move cautiously in assessing the risks and put in place effective safeguards against hackers and system errors. All this means jobs in banking and finance are changing, and moving toward the technology. As referenced earlier, financial firms struggle to attract software engineers with finance expertise. That's being addressed. The New York Institute of Finance and Google in 2020 began offering a course in machine learning for trading specialization.[16] The CFA Institute provides advice on becoming a data scientist in the investment industry. It highlights big data, statistical analysis, data visualization, and industry-specific data analytics as key skills to learn.[17] While regulations, extra levels of security associated with finance, and inherent institutional caution have caused banks to go slower than technology companies or hedge funds in adopting and scaling machine learning systems, now that the technology is more established, they are going at full speed ahead.

Chapter 18

Self-Drivers

Cars and trucks are starting to drive themselves, but, as we examine in the following, can they cope with the many perils of the road? Because of the recent improvements in sensors, self-driving vehicles are using machine-learning technology to solve problems that were previously thought impossible. There are still edge cases: Errors, and crashes, may come from unexpected places. With the advent of self-driving cars and self-driving trucks, we also discuss the prospects for self-sailing boats or self-flying planes.

In 1987, Ernst Dickmanns, a German engineer, created the world's first test of an autonomous vehicle on the German autobahn. More than 35 years later, widespread adoption of self-driving cars is still far from reality. Autonomous vehicles always seem to be on the cusp of being realized on our roads yet in a state not quite safe or reliable enough and always "stuck in perpetual testing."[1] Full self-driving cars have taken far longer to come to fruition than predicted by technologists and the car industry. The main reason for this is that driving is a much harder task for computers to handle autonomously than other kinds of activity. You can see why, having seen machine learning in action in other instances, where it has a far narrower band of focus. Driving isn't just one simple procedure but a multitude of concurrent tasks in vision, perception, and action. While humans can learn to drive fairly easily, it's actually a highly complex operation, with unexpected conditions, multiple permutations, and combinations of possible outcomes. This harks back to Steve Wozniak's coffee challenge. Weather changes, unpredictable motorists, pedestrians, equestrians, and events that simply haven't been seen before make autonomous vehicles hard to perfect. There are also the raised stakes of safety and human life. It's one matter for the word recall accuracy rate to be subpar in a machine translation model, but mistakes on the road can be deadly and are a different matter entirely.

Baby Robot Drive My Car

While the difficulty is great, the payoff is high. Few people like driving for its own sake, and there's the potential to save millions of lives as road traffic deaths are so numerous. So companies and researchers have plowed ahead. While computer scientists initially underestimated the complexities of teaching cars to drive, we are now at a stage where self-driving cars are available and working, if not perfectly. But it turned out to be harder than expected to get them on the roads and is taking far longer. Elon Musk has been promising full autonomy in Teslas for at least 5 years. There's also the realization, as with elsewhere in industry, that immediate full automation may not be the best approach, certainly to start with as the technology and safety need ironing out and iterating. Better is a system of enhancement that takes much of the burden away from the people, while leaving a measure of control. (Similar to the Human on the Loop idea in the military.) And this is a convenient stop-gap measure, even if full automation is the ultimate goal. Bad weather and unexpected events or conditions can also derail or upset autonomous vehicles. There are so many variables in driving that humans can just automatically deal with, which cause computers to freeze or crash if they haven't been explicitly trained to handle them.

Autonomous vehicles are usually categorized into five levels[2] (with an additional "level zero" consisting of no automation at all).[3] Stage one is driver assistance, where the human remains in control and a single automated system helps out, such as cruise control. Level two is additional assistance, where the driver is in control with continuous help, and the car can control steering and acceleration. Level three is known as conditional automation, where the car is in control with the driver able to step in at any time if the system fails. Level four consists of high automation, which means no human is needed, but the vehicle operates in limited areas. Finally, in level five, the car is fully in control in all conditions and on all roads. Humans aren't needed.[4]

Tesla's Mountain of Data

Tesla is the market leader in electric vehicles and has a big stake in the future of self-driving cars as it designs its cars and software to upgrade to autonomy. Musk has been consistently overoptimistic in his predictions about when Tesla will achieve fully automated vehicles. Tesla collects a mountain of data from its fleet of vehicles that it uploads into the cloud and uses to improve safety and head toward autonomy. Its fleet of more than 1 million vehicles beams back data in real time. Tesla believes that artificial intelligence is crucial to achieve full self-driving cars, specifically in terms of vision systems. Here's what the company says: "An approach based on advanced AI for vision and planning, supported by efficient use of inference hardware, is the only way to achieve a general solution for full self-driving and beyond."[5]

These cars have a boggling array of sensors: cameras, weather sensors, and steering wheel sensors. Each Tesla contains a dashboard pad with a real-time image of the road around it, including pedestrians, other road users, and a map. This is all aided by machine learning. Tesla uses neural networks to work on problems of perception and control. The company uses camera images to work object detection, semantics, and depth estimation. Tesla also uses a bird's eye view to gauge road layouts, static infrastructure, and three-dimensional objects. A "full build" of an autopilot network demands 48 networks and 70,000 GPU hours to train, Tesla says. They produce 1,000 predictions in each timestep (a kind of way of measuring data over a specified time interval). And Tesla says its autopilot system can cut accidents[6] by 40%.[7]

BMW and Daimler are among traditional carmakers also racing to adopt machine-learning technology. BMW has a twin focus to streamline and automate its business processes and add smart technology to create better products.[8] In Munich, BMW built what it describes as the world's most advanced driving simulation center, letting it gather data for autonomous vehicles faster. It has also built some concept models that showcase cars assisted by machine learning. BMW is using computer vision technology from Intel and driving data technology from IBM. Daimler's Mercedes unit is using artificial intelligence to make production and transportation more efficient. Many self-driving projects are still in the pilot phase as of this writing. Daimler has built what it called the first self-driving heavy goods vehicle. And Daimler partnered with Nvidia to design deep learning systems, so it is certainly focused on a future of autonomous vehicles driven by machine learning.[9] The way self-driving cars and trucks work is in three parts: perception, prediction, and driving policy.[10] Sensors include cameras, radar, and Lidar. Once the data are collected, the car needs to figure out its surrounding objects, including pedestrians, bicycles, horses, road markings, obstacles, and more. Things that have confused the software range from puddles in the road, to bicycles on racks on the backs of cars, to a man dressed as a chicken.[11]

Three Zeros Pledge

So breakthroughs are still needed for full autonomy. And self-driving cars won't be widely deployed until they're consistently safer than human drivers. But car companies are sticking with the vision. Mary Barra, chief executive of General Motors, has pledged a goal of "zero crashes, zero emissions, and zero congestion."[12] Elon Musk certainly sees self-driving technology as crucial to the future of Tesla. In terms of autonomous vehicles, cars are just one path. There are also trucks, which haul our goods from one end of the country to the other, and container ships, which do the same thing across the sea. Self-driving ships and trucks are also in the works. The US government announced a $100 million spending bill to help automated vehicles overall and a plan of action.[13] Driverless vehicles hold promise for transportation safety, society, and the consumer, according to a congressional hearing in 2017. A patchwork of different rules and laws across states may raise the barrier to entry. Policy makers should foster the industry while focusing on lowering risks to drivers and passengers.

Autonomous vehicles also have additional ethical considerations beyond other algorithms.

Consider the decisions cars must make in an emergency where the system needs to decide between injuring or even killing a pedestrian, or risking its driver. Do children passengers take priority over adult pedestrians? These decisions must be considered by the algorithmic designers, and they aren't even easy for humans to make. Should cars have built-in accident decisions, and if so, which ones?[14] Perhaps it would have been better for the industry to start with a push toward self-driving trains, which by nature have a more predefined route and far fewer parameters or permutations than a car. There are many real-world examples of trains driving autonomously for years, for example, airport terminal shuttles and some metro routes. The first fully autonomous driverless train network was Port Island Line in Kobe, Japan, which has been operating since the 1980s.[15]

Trucking Along

As for self-driving trucks, the same problems and hurdles in the development are occurring as for cars. Ford is rolling out an electric version of the F150, its most popular truck, which includes hands-free driving. Rivian is making electric trucks and SUVs with hands-free functionality. While these technologies aren't fully autonomous, they are a step that way, and both companies likely have autonomy in mind. Waymo, owned by Alphabet and formerly the Google self-driving car project, is building what it calls the world's most experienced driver. The company's mission is to make moving people and things safe and easy, presumably all with self-driving technology. Waymo has integrated a series of sensors and computers designed to perceive its surroundings.[16] The company is working toward full autonomy for both cars for passengers and trucks to move freight. Volvo in 2021 unveiled a long-haul truck as a prototype for use in North America, a step toward fully autonomous Class 8 trucks, which haul heavy goods from coast to coast. It also launched an autonomous bus in Singapore.[17]

Boating on Autopilot

Autonomous boats are also becoming a reality — ships that steer themselves. In Norway, a battery-powered ship is ferrying fertilizer from factory to port, following six months of tests, trials, and outfitting.[18] Yara Birkeland is the first electric autonomous container ship. MIT's Computer Science and Artificial Intelligence Laboratory (CSAIL) and the Senseable City Lab launched robot taxis for the city of Amsterdam in the Netherlands, after a 5-year project to build the first fleet of autonomous boats.[19] They are big enough to carry five people and also could be used to transport waste. They are battery-powered and use Lidar and cameras to survey their surroundings.[20] The arguments of road safety and lowering traffic deaths combined with the need to reduce emissions to combat climate change are compelling enough for companies to persevere with self-driving cars and trucks, despite the obstacles and setbacks, along with the technical difficulties in teaching machines to drive alone. The same holds for boats and trains, as long as they can be made to operate in total safety. Autonomous vehicles are the future of transport; it's just that the rollout might include more human input than previously thought.

Chapter 19

Don't Believe the Hyperdrive

A multitude of companies now say they are harnessing or leveraging artificial intelligence. Often, it's exaggeration or just plain data wrangling. How do we cut through this hype to understand who's doing what, how, and where? And where is the real cutting-edge work taking place? Companies are keen to tell customers they're leveraging the latest AI technology. In this chapter, we evaluate if all the hype presages a bubble that will burst or if the underlying change is transformative enough for society to avoid a giant pop.

Deep learning and artificial intelligence are hot topics, so every company now is jumping on the bandwagon to say they have some machine-learning function, specialized department, or expertise. These are often just rudimentary data analysis or data wrangling teams. Even if companies have data scientists, it's still hard to integrate the majority of such theoretical work into practical business systems and tailor them for specific business needs. But companies must strive to show investors and customers that they are still at the cutting edge of their industry. So how do you show that machine learning can be useful and have a clear business use? And how do you ramp up the systems to show you're at the sharp end of your business? That's what many companies must now deal with, not just in the technology industry. As so many companies are terming themselves AI First, there's the question of how this really manifests and how to achieve it. What does all this marketing mean? There's a parallel with the dot-com boom and bust in the 1990s and early 2000s, where companies raced to rename themselves to include dot-com. There were also the 2022 layoffs in the technology industry where executives hired too quickly for perceived growth that didn't materialize, and this affected the technology giants as well as tech startups. Companies now want to be associated with artificial intelligence whether they legitimately use it or not. And artificial intelligence technology is moving so quickly that the real cutting-edge technology companies are developing off-the-shelf solutions that smaller companies can use without needing hundreds of machine-learning engineers to build. The corresponding jump in computer processing speeds and cloud-based systems means that those tools can be used without the huge data architecture of yore.

Intelligent Design

Google, Amazon, Microsoft, and Facebook are building machine-learning platforms that are looking to the future. Their design decisions, in both hardware and software, say a lot about the future of technology, and decisions made now will be key for the outlook in the coming years, another reason why ethics and education are important, along with careful algorithmic design. Technology giants like these are less susceptible to hype because of the sheer amount of data at their disposal and the fact that their leaders have explicitly pivoted to become "AI First" companies. Google's Sundar Pichai adopted such a strategy and Satya Nadella pivoted Microsoft to AI over several years. Mark Zuckerberg detailed how he built an AI system into his home to control things and is pushing Facebook toward the metaverse (which will likely need machine learning and more to capture users and adoption). Amazon's AWS relies on offering machine-learning infrastructure to other companies, showing their own somewhat meta AI-first strategy from 2016 and before. Similar arguments hold with Microsoft's Azure. When it comes to companies that don't have petabytes of data at their fingertips, however, it's less clear where the cutting edge stops and the waffle begins.

Puff Versus Reality

While machine learning has faced cycles of boom and bust before, where the reality didn't live up to expectations, this time the technology is infusing every part of our lives. From the tax software you use and the ride service you take to the search engine you favor, companies are increasingly relying on machine learning for a greater part of their business. That's not to say there's no skepticism out there. Some experts argue that we're in a bubble that will pop when expectations fail to match reality, with job cuts as evidence of that. Others say that the technology evangelists are over-promising as in the past, and disappointment will ensue:

> *Despite this history of missed milestones, the rhetoric about AI remains almost messianic, say Gary Marcus and Earnest Davis. The bitter truth is that for now the vast majority of dollars invested in AI are going toward solutions that are brittle, cryptic, and too unreliable to be used in high-stakes problems.*[1]

Hype and reality are also important when distinguishing how companies are looking to adopt systems to fit their business or implement a specific use case. Knowing when to use a machine-learning model or when it's appropriate to leverage a learning system for a set task is crucial for future-proofing businesses as well as avoiding costly blunders. Machine-learning engineers are expensive, and building systems that produce results in line with business goals is hard. A lot of companies claim to be using machine learning when what they are doing is simple data analysis. There's also confusion about what can be achieved by different systems. Sometimes, a problem or task doesn't need complex machine learning and can be solved with a cleaner database query or some business analytics. That's not to say there's anything wrong with simple data analysis that isn't based on the latest machine-learning model. Often, business needs are better met by clever use of an SQL database or some clear visualizations showing business performance. But artificial intelligence is such a vague term that organizations are able to boast of futuristic usage that isn't backed by reality. Data scientists and data architects will hopefully be able to bring a renewed rigor to industry such that they can explain to executives exactly what is feasible for their companies, what the data strategies should be, and how to go about it.

Skills in Demand

All this presents opportunities for employees at the periphery of such systems. Those who can work with computer scientists and data engineers and bridge the gap with executives who have heard the latest buzzwords, but don't understand the nuances, are at a major advantage. If you can cut through the noise, understand business needs, and translate technical terminology into clear language that senior managers and executives can understand, you are in good stead. Hiring the best people in semi-technical roles such as project manager, product manager, user interface expert, and industrial designer will become key for companies. These are the jobs that will determine the success of a given project as much as the machine-learning engineers and coders who build the underlying systems.

Nitpicking over what is and isn't machine learning is unnecessary as all companies strive to become more data driven and to showcase the best of their technology. But it's still really key for executives to pick the right technology for the right job. Building expensive white elephants in machine learning is a real risk for companies that don't know the precise nature and scope of their demands. Managers must have a strong handle on the capabilities of artificial intelligence in its various forms, including its strengths and weaknesses, and where it applies to their business. InfoWorld points out that machine learning is going through industrialization and needs to have real business applications:

> We must resist being swayed into believing that cool and fanciful machine learning demonstrations — such as writing poetry and generating clever dialogue in video games — are the norm or the path forward for machine learning in the real world.[2]

Machine learning must solve complex business problems, provide actionable insights in real time, and get integrated into operational processes to be truly useful. This is harder than it seems. Software engineers also have a propensity to solve the most difficult problems, meaning scope creep that causes projects to be delayed and costs to spiral without careful understanding and management of the goals and aims. Also machine-learning models must be picked carefully to suit the problem they are solving, or a lot of wasted time and energy can ensue, along with dollars down the drain. In terms of the whole industry, some worry that hype is at a stage where we are in a machine-learning bubble, and an AI Winter will necessarily ensue. But there have already been so many provable successes in the past decade, in industry as well as academia, that an AI Winter is by no means a given.

Venture capital investment is one measurable indicator of the future of the industry, as is capital spending by public companies in machine learning. Private investment in artificial intelligence globally fell for the first time in a decade in 2022, to $91.9 billion, according to the AI Index Report from the Stanford Institute for Human-Centered Artificial Intelligence. Still, investment has surged over the ten years, climbing by 18 times since 2013.[3] Policy-maker interest in AI has increased, and industry has raced ahead of academia. Further, demand for professional skills related to AI is growing across almost every part of American industry, the report found. In the future, companies will need to assess if they need to build their own machine-learning teams and systems to keep their business competitive or buy off-the-shelf solutions from a technology giant or a niche player. The answer will depend on the kind of tasks and business unit involved, along with the industry. This assessment in itself will yield job opportunities for advisory or consultant work that can offer this kind of service. The smart movers will be a step ahead of the competition in this realm. Also key will be avoiding the pitfalls of building expensive systems that are soon defunct or not fit for purpose.

Data architecture and systems design are crucial when thinking through these business questions. The reason so many technology giants retooled to become AI-first is that they predicted the technology will rely on learning models and systems that self-improve in the future. They also predicted the power of deep learning and neural networks when combined with big data and increasing computing resources. Sometimes, a complex machine-learning model isn't what is needed to address a business need or solve a problem. But many companies need data analytics to see how they are doing in a granular and timely manner, forecast trends, and keep up with their customers and vendors. This kind of work can be achieved with data, but often database interrogation combined with useful dashboards to showcase the results is what is needed. Identifying the right tools and the right technology can help the future direction of the business. Machine learning isn't useful for everything, as the skeptics are quick to point out. It's also expensive to build and needs maintenance. That means finding a chief technology officer with a strong handle on which technologies are key to achieve what business need. Mistakes will be made.

Chapter 20

Adversarial Art

In the following pages, we cover the weird and wacky uses of technology within the art world. Artists treat machine learning with a refreshing irreverence and playfulness compared to people in other industries. Is artificial intelligence a tool or a medium, or both? How are Generative Adversarial Networks (GANs) used in the latest art exhibitions? As algorithmic models are becoming more commonplace in artists' studios and galleries, can machines create art? Such questions are deeper than they seem and raise issues related to ownership, prices, and hype.

B otto is an autonomous robot artist that generates art using a machine-learning system. In late 2021, its first six works sold at auction for around $1.3 million. Botto self-educates. The robot generates about 350 pieces of art a week and then seeks votes from its community on which pieces are the best, which it then uses to train its algorithm and adapt the kind of art it creates. Botto was initially formed by a group of engineers based on an idea from German artist Mario Klingemann.[1] Artists and musicians are adopting machine learning technologies at an explosive rate. But their use is at once more playful and irreverent than elsewhere in industry or in finance. They also have a different purpose. Here's artist Sofian Audry on the subject:

In contrast to computer engineers and scientists, these artists do not use machine learning because of its precision but for its open-endedness. They are not so much interested in playing the imitation game of computational creativity as they are in repurposing machine learning algorithms in ways for which they were not initially designed.[2]

Audry, also a computer scientist, says the "frenzied enthusiasm" from artists is accompanied by myths and misconceptions about artificial intelligence and machine learning, which he aims to debunk. New media arts have also been overlooked by art historians, a gap that needs addressing. Audry uses machine learning in his own artwork. Learning systems have become a new material for artists, and as such, artists have an important role in addressing the technologies beyond their scientific and commercial uses, Audry says. He's right.

There are also some more practical uses of machine learning in the art world. Sifting through vast catalogs and making comparisons is a perfect task for machines. New York's Museum of Modern Art teamed up with Google to categorize art from 30,000 exhibition photos. MOMA's Digital Media team and the Google Arts & Culture Lab sought matches between the exhibition photos and the thousands of works in MOMA's online collection. The resulting identification of around 20,000 works provided a network of new links between the museum's exhibition history and the online collection, boosting a resource for museum visitors and improving its databases:

Now a photo from a 1929 painting exhibition opens a window into an iconic work by Paul Cézanne; a 1965 shot of Robert Rauschenberg prints connects you to those same works in MoMA's 2017 Rauschenberg retrospective; and one corner of a 2013 design exhibition becomes a portal into poster art across two centuries.[3]

MOMA described the project as a "remarkable feat" given the volume of information. Meanwhile, there's something otherworldly about Botto's art. It has an amalgam feel to it. But it would be hard to pinpoint it as computer generated without prior knowledge or unless you are an expert.[4] It's not just visual arts where machine learning is taking hold.

Making Machine Music

Rebecca Fiebrink is a London-based computer science professor who teaches a course in machine learning for artists and musicians. She built a special tool for creatives to harness learning technology to build new musical instruments using gesture control. Called Wekinator, it takes a variety of inputs and processes them to control outputs. Musicians can use it to make new instruments, connect gestures to sounds, analyze gestures, and detect pitch or rhythm. The software can also be used to make animations or games controlled by gestures. The system uses supervised machine-learning algorithms and is designed for the interactability needed by musicians, the movements and interface as the feel of three-dimensional gaming.

GANs

Generative Adversarial Networks, or GANs, are a type of machine learning that trains two models at once and where the two compete to be the most accurate. These models have taken off in art because they can generate pictures in the likeness of artists. They have also been used to generate "deep fakes," pictures, or videos in the likeness of a real person generated wholly by computer. GANs are generative, meaning that they build new art with data points based on training data. GANs consist of two distinct neural networks, known as the generator and the discriminator. These act like a counterfeiter and policeman, with the counterfeiter constantly improving its fakes and the policemen trying to catch it out. Generators take in data, such as an image, and mimic that dataset. Discriminators try to assess if the data are a real input or a generated fake. GANs can take large datasets and, via unsupervised learning, figure out some patterns and create new images that look like the originals. "Different species of machine learning systems afford different kinds of artistic practices and aesthetic qualities," says Audry.[5] The Camera Culture group at MIT Medialab is exploring different platforms, across sensors, health, automating machine learning, and geomaps. One project is to help cameras see around corners, beyond the line of sight, or even cameras that can see through skin: "Like the invention of applied pigments, the printing press, photography, and computers, we believe machine intelligence is an innovation that will profoundly affect art," says Blaise Aguera y Arcas, of the Artists + Machine Intelligence project at Google. "Systematically experimenting with what neural-like systems can generate gives us a new tool to investigate nature, culture, ideas, perception, and the workings of our own minds."[6]

Dall-E 2 is a giant system that generates art from phrases in natural language, built by OpenAI, the maker of ChatGPT. It works by automatically creating images and art from a description in sentences given by the user. In other words, you type in the art you want to see, and Dall-E 2 creates it instantly. It can draw from a text description, edit existing images, and make variations inspired by original art. It's come up with some weird and wacky images on demand, but the key is that the system has figured out how to pair different concepts together, allowing it to draw a fish on a bicycle, for example.[7] The platform, whose name is a play on Disney robot Wall-E and surrealist artist Salvador Dali, uses a process called diffusion, where it takes a pattern of random dots and turns it into an image, as it recognizes parts of the picture. Dall-E 2 was built by training a neural network on pictures using their text descriptions. It not only detects objects but also the relationships between them. It has some limitations. If its original images were incorrectly labeled, these errors will be replicated, for example, if an image of a car was wrongly labeled "plane," when you type car, it will draw a plane. Also, it suffers from gaps in knowledge if certain data is absent from the database. The system also contains some safety measures, including preventing violent, hateful, or adult images. Dall-E 2 has checks to curb misuse and a phased rollout limiting it to trusted users.

Uncanny Valley

Artificial intelligence is popping up all across the art world in unusual ways. The Uncanny Valley is the phrase used to evoke the feeling of unease at human-like robots. An exhibition in San Francisco, "Uncanny Valley: Being Human in the Age of AI," explored current applications and challenged traditional human–AI relationships. Fourteen artists made works to show how artificial intelligence is incorporated across all aspects of life. New York artist Ian Cheng built an artificial "lifeform" called Bag of Beliefs, or BOB, which took the form of an animated branching serpent. BOB learned from its interactions with viewers. Via an iPhone app, it was possible to visit BOB's shrine, to send virtual gifts, upon which BOB's behavior would change. The shrine automated stimuli for BOB based on the users, and BOB responded by handing out rewards in reciprocation:

> BOB occupies a strange place in the landscape of artificial intelligence systems. Unlike deep-learning systems that require thousands of data points, BOB can learn, unsupervised, from few examples, the artist says.[8]

Digital artist Beeple,[9] also known as Mike Winkelmann, was surprised and amazed by the responses after asking people on X (previously, Twitter) to recreate their versions of his work using artificial intelligence. The public responded to the challenge with a wild array of machine-created art. Unlike office workers, Beeple seems more delighted than fearful of the technology.[10]

Is It Art?

Is art made by machine really art? Artists and critics are divided on the answer, with many artists saying that machine learning and artificial intelligence just represent new tools of expression. Others say that a model that imbibes thousands of paintings from history and mimics their various styles to produce new work is merely copying, rather than generating anything new or truly creative. At Rutgers University's Art and Artificial Intelligence Laboratory, researchers built a Creative Adversarial Network, instead of a Generative Adversarial Network.[11] The idea was to make something more creative, as GANs ultimately make something that typically looks like existing art.[12] The AICAN was designed to make art that didn't follow established styles.[13] AICAN now operates as an autonomous artist, similar to Botto. Its work was shown at exhibitions in Frankfurt, Los Angeles, New York City, and San Francisco, with a different set of images at each, and the work was credited to the algorithm rather than the computer science researchers who built it.[14]

Ownership and Rights

Where do the credit and ownership lie if a machine creates a piece of art? It seems unlikely that a machine can hold the copyright of a work of art, but it might be held to be the author, at least in some countries. These questions are still to be determined and represent a legal gray area as both the technology develops to allow more artistic autonomy, and the legal frameworks in different jurisdictions evolve to handle such nuances. Some artists explore the darker side of AI. Works are starting to examine machine bias and automation eliminating jobs. There are also some bigger metaphysical concerns, in terms of the forces we are unleashing on the world that mirror philosophy's existential worries about the technology. Neuroscientist and poet Abhijit Naskar wrote a sonnet called "Humanizing AI (The Sonnet) | Either Reformist or Terrorist" to explore concepts in AI through poetry.[15]

Art Outlook

The future looks bright for many artists, as they are embracing the technologies but finding things to say that challenge and criticize the state of the world as it is. Artists seem less worried than other workers about machines taking their jobs, even though computer systems are getting better and better at art. Machine learning is also providing fertile ground from which artists can draw both as a source of commentary on modern technology as it pervades daily life and as a new medium through which to express themselves. As the technology accelerates and science veers toward yesterday's science fiction, artists will be at the forefront of people with meaningful things to say about the state of affairs in the future of artificial intelligence, as they have no need to evangelize the technologies and approach the subject from a different perspective. They also exhibit a certain enviable lightness about the whole subject.

Chapter 21

Education, Education, Education

Education needs to change to furnish the workplaces of the future with much-needed highly skilled staff and for society as a whole as machine learning becomes more and more commonplace. Why are schools falling short in teaching this technology? How does machine learning have the capacity to provide tailored education services? Moreover, cheating is becoming a growing concern for teachers and parents. In this chapter, we explore if this technology can revolutionize teaching or if it is a pipedream.

Imagine having a personal tutor with the world's knowledge at his or her disposal and with whom you can spend unlimited time. That's one proposition that machine learning offers the future of education. Adoption in education is at an early stage compared with many other industries, but in all likelihood, it will have an outsized impact on the sector because education is so crucial to society's future. This is especially true where it facilitates classes and lessons tailored to the individual. The idea of customizing education to fit each student at scale would be impossible using individual teachers; there simply isn't the manpower. Using machine learning techniques it's theoretically possible to create a whole curriculum for just one student or to create user-specific lessons for each subject, paced correctly and aimed at the right level to ensure maximum success, a kind of one-on-one tuition, but using machines. AI pioneer Kai-Fu Lee makes this prediction:

> With AI taking over significant aspects of education, basic costs will be lowered, which will allow more people to access education. It can truly equalize education by liberating course content and top teachers from the confines of elite institutions and delivering AI teachers that have near zero marginal cost. I believe this symbiotic and flexible new education model can dramatically improve accessibility of education.[1]

So machine learning has the potential to have one of its biggest impacts in the education sphere if it can achieve such customization and tailored experience at low cost.

Learn to Address the Skills Gap

Spending is increasing. Artificial intelligence-assisted education will be worth at least $5.8 billion by 2025 and far higher in later years, research from the Association for the Advancement of Computing in Education suggests.[2] As society becomes increasingly reliant on data and data-bases, children and young adults need the skills to understand, handle, and navigate the data around them. Education needs to adjust to that. Educators are changing pedagogy and teaching methods to take this into account but too slowly. Employers complain of a skills gap when it comes to new technologies and changes taking hold in the workplace. That's an added incentive for schools to address the latest trends and adapt to equip students with the tools they need. There are two aspects to machine learning within education. First is the harnessing of the technology for students, teachers, and professors to use. Second is the subject matter itself. Students and teachers need to be trained in understanding and controlling this new technology. (As mentioned earlier, consider the flap about OpenAI's ChatGPT being used to cheat on tests. This kind of thing needs to be thought through by educators and quickly.) Companies are already lamenting a skills shortage for technology and engineering roles.

And data scientists are in demand. The role of education in providing students with these skills is key. Further, the periphery jobs surrounding the technology sector will also need related key skills and knowledge, even if not hardcore coding abilities. As discussed in other chapters, computer science education also needs to have ethics and safety baked into education and training at an early stage to avoid design problems later regarding bias, transparency, and security. Education is where it all starts.

MOOCs and More

Technology is already making education more widely available at lower cost through Massive Online Open Courses, or MOOCs. Leveraging machine learning is the next step to democratize education. Coursera is one example. Founded by AI pioneer Andrew Ng (also of the Google AI cat detector), Coursera works by linking up with more than 250 universities and companies to provide relevant training and skills online.[3] There are a mixture of free and paid courses as well as professional certificates and degrees. Ng's own course on machine learning, the first taught on Coursera, is highly popular, and he updated the content in 2022.[4] Six key ways that AI will revolutionize education are through learning analytics, predictive systems, automated assessment, increased efficiency, personalization, and adaptive learning.[5] Teachers are starting to use machine-learning technology to watch for students who are struggling to keep up, in order to act to help retention and boost success. Students already can use it for transcription, turning text to speech, and personalization. Professors are using it to enhance enrollment and predict outcomes, and researchers are using it as a tool to gain new insights. According to Amazon, some of these tools are already in place:

> *Educators are using ML to spot struggling students earlier and take action to improve success and retention. Researchers are accelerating research with ML to unlock new discoveries and insights. ML is expanding the reach and impact of online learning content through localization, transcription, text-to-speech, and personalization.*[6]

The role of teachers and professors is changing. Automation and AI can help with administrative tasks, creating educational content, personalization, and virtual learning.[7] One project used training data from Ugandan students to analyze performance. The average class size there is 114 students, so one-on-one tuition is impractical at best. This model detects "wheel spinning" students who need help. The system was able to predict interventions 80% of the time in simulations. In a real-world setting, the model mostly agreed with experts but with some limitations and some errors in the model inputs. Another study looked at a data analysis tool developed at the University of Waikato in New Zealand to monitor student performance in Oman, devise preventive measures at an early stage in the semester, and offer help to struggling students. The study compared a variety of models for the best outcome and found decision tree systems worked the best. They also wanted to turn the results into a usable form for teachers to view findings and implement changes.[8]

It's not all plain sailing. Using algorithms and machine learning unthinkingly in education, as elsewhere, could have unintended consequences and even exacerbate bias and inequality. Machines are no substitute for good teachers. There's also the risk of technology addiction as students spend more and more time on devices anyway. That, combined with their time on social media outside school hours, can increase stress. Computer scientist and journalism professor Meredith Broussard points out that data-first approaches and solutions that rely 100% on technology don't necessarily work for social problems. And this is evident in education. She uses the example of Philadelphia public schools to show how a shortage of books makes it harder for students in these schools to succeed at standardized tests. Broussard argues that standardized tests in US education will never succeed because of the way that they are set and the way that tests rely on textbooks that aren't available. More computers or electronic aids won't help. Data aren't immutable truth, and data collection systems were created by people, she says.[9]

It's naive to believe that data alone can solve social problems. The Philadelphia schools don't just have a textbook problem; they have a data problem — which is actually a people problem.

Another data problem lies in opaque algorithmic school rankings. Cathy O'Neil, who coined the term Weapons of Math Destruction, points out that algorithms can have a negative impact in terms of how colleges are ranked because of a "vicious feedback loop" because of self-reinforcing rankings.[10] If a college fared badly, its reputation would suffer and conditions deteriorate, and top professors and students would shun it.

Self-learning is gaining popularity, and barriers to entry are getting lower. Still, MOOCs have great scope to expand to developing countries and places where internet access is more limited. Online courses are unlikely to replace undergraduate education. They're more likely to be a supplement or a way of professional upskilling or retraining. So-called Intelligent Tutoring Systems can provide tailored and granular feedback to pupils. Some test companies use natural language processing to grade essays. Early warning systems, which act similarly to recommendation engines, can help monitor struggling students or predict failing academic performance.[11] Vision systems can also be used, for example, a teacher can take a photo of a mathematics equation and a machine can automatically grade it. Discovery Education partnered with Amazon Web Services to use machine learning to help teachers sift resources and get personalized recommendations for K-12 learners.[12] Microsoft has a free tool that helps teachers to track progress in the fluency of students' reading.

Costs and Benefits

Artificial intelligence systems help accelerate learning most when paired with strong teachers and top-rated books and materials.[13] They can help teachers identify struggling students earlier. Other specific ways the technology can help include the automation of classroom tasks, creating smart content, course personalization, virtual learning, chatbots offering 24-hour assistance, facial recognition, and virtual exams.[14] Privacy concerns are always an issue in education (as with other industries that necessitate sensitive datasets, such as in healthcare). How can systems that collect student data in the classroom guarantee privacy, and how can the data be anonymized while remaining useful? These questions need to be thought through by educators before adopting such technology. Also, schools will need consent from both parents and students. Machines can't replace teachers. Risks of bias are always inherent in the systems, especially when it comes to testing. And there's also a need to explore the cultural and social impacts of AI in education.[15]

Embed Good Practice

In future, all computer science education should include embedded training on ethics to teach students about algorithmic bias, the pitfalls of machine learning systems, and all the ethical issues that arise as the pace of technological change accelerates. That will make sure the future workforce is correctly trained on all the necessary aspects, not just the technical side of the discipline. Education needs increased investment to keep the technology at the cutting edge. Schools, colleges, and universities should focus on systems that combine machine learning technology with the best of their teachers with the top books and materials. New technologies should enhance existing systems and processes rather than replace them.

Chapter 22

Social Networking

In this chapter, we evaluate how social media algorithms make suggestions and handle friend requests, how machine learning is helping social networks connect your data and link with your connections, the growing risks of fake news and spam in social media driven by algorithms, and how such mechanisms work in the real world at Facebook, LinkedIn, Instagram, TikTok, and X (previously, Twitter). We also explore how dating apps use machine learning to help with matching, and the key swiping interface on phones.

Most people on social media platforms have no idea that artificial intelligence is driving so much of their interactions, from engagement, to recommendations, to newsfeeds. Social networks are a natural fit for machine learning. Big data, cutting-edge technology, and engaged users all add up to a perfect fertile ground for data experimentation. Social media companies have millions or sometimes billions of users, interacting and engaging with the platforms, uploading content, making comments, and adding connections. Machine learning lets social networks divine their users' emotions, likes, dislikes, and predilections. This lets them build a sophisticated view of user behavior and therefore provide laser-focused targets for advertisers and, hopefully, more useful ads for the user. So it's no surprise that Meta is at the forefront of using such technologies, both via Facebook and Instagram. Also, the stakes are lower in social media than in healthcare or self-driving cars, where lives are at risk. That means experimentation and innovation are easier.

The backbone of machine learning at Facebook is a system called FBLearner Flow. Machine learning is used to personalize experiences, rank and prioritize peoples' newsfeeds in the platform, and filter offensive posts. It also points out trending topics and ranks search results. Facebook retooled its internal architecture in 2014 to make it easier for all Facebook engineers to take advantage of its machine learning architecture. Before this, the tools were hard to access for all the staff. This is a trend taking place across other companies too, as machine learning becomes more ubiquitous.[1]

Monetize Data

With 3 billion users, social media companies are in a prime spot to suck up data and monetize it. They can also use it to learn a lot about their users' preferences. All this is gold to advertisers and marketers, evidenced by spending on social media advertising and the decline in ad spend in traditional media. Conversely, users have shifted their behavior on social media to become consumers, hunting products, browsing and purchasing, and making customer service inquiries. Companies are using artificial intelligence for "Social Listening" to get a better understanding of customer interests and behavior, and monitor "Brand Mentions" for insights across social platforms.[2] Social media can make use of sentiment analysis to understand users' viewpoints, along with text recognition, image recognition, and content recommendations. Chatbots are also on the increase.

Hate Speech and Fake News

There's a downside to social media platforms. Hate speech, cyberbullying, and fake news are all extremely hard to prevent at scale. And many social platforms have argued that they aren't media companies, therefore they aren't responsible for the content posted on their systems. This has become polarizing. Mark Zuckerberg's view that artificial intelligence should be able to solve such problems by detecting and categorizing unacceptable content has proved hard to implement effectively. For example, cleaning up hate speech and fake news turned out to be much more nuanced than the company thought. Free speech arguments also play a role, with conceptual disagreement about what is acceptable on any one platform. X (previously, Twitter) is the posterchild example of social media platforms facing thorny questions surrounding free speech. Elon Musk's purchase of X and the subsequent fallout only emphasized that. But other platforms face many of the same problems. Deleting or erasing harmful content isn't always the answer.[3]

Meanwhile, Instagram, also owned by Facebook parent Meta, is using machine learning to combat cyberbullying.[4] It works by training models to recognize offensive comments or spam and automatically block them. The system also improves over time, incorporating text, photo, and video. Instagram started with a filter for spam and toxic comments and moved to include the prevention of bullying, such as attacks on people's character or appearance, or threats. One feature uses an artificial intelligence system to notify people that a post may be offensive before they publish it to the platform.[5] It's not perfect, as parents can attest, but it shows the way the companies are using such systems to address concerns.[6] Facebook is using similar technology to monitor users who may be depressed or suicidal on its platform. The system matches patterns in user behavior to previous posts where people were at risk of self-harm.[7] In general, Instagram uses a different algorithm for each part of its app, customized to how each function is used: a user's feed, their stories, their reels, along with explore recommendations, and search functionality. Stories are geared toward people's closest friends, whereas explore is designed for discovering new content, and reels for entertainment. Instagram uses different rankings for each, accordingly.

Social media and machine learning go hand in hand because the network effects allow platforms to leverage their user data to predict and suggest things to their users in a self-reinforcing way, that both helps keep users on the platforms and serves up (in theory) constantly improving content. At LinkedIn, owned by Microsoft, there's a gigantic interconnected dataset that's focused on the professional world. That's in contrast to Facebook, Instagram, and other social media apps that started with people's personal lives and now sell data to businesses to advertise to users. LinkedIn uses machine learning to match companies with prospective employees. It also uses the technology to help with suggestions, connections, and learning. LinkedIn has said all its engineers will eventually be trained in how to use artificial intelligence,[8] yet another example of companies striving to be AI-first. Using learning models for job matching may also unearth candidates who might otherwise be overlooked. LinkedIn tries to build models that accurately predict what companies are looking for during recruitment. LinkedIn uses knowledge-graph technology, which builds a kind of multi-dimensional map by linking entities, such as members, jobs, titles, and companies to form relationships and allow complex database lookups.[9] In one sign that is working, revenue has jumped since Microsoft bought the company.[10]

TikTok, owned by Chinese company ByteDance, has made great inroads into attracting millennials and Generation Z youth with its brand of short videos and its addictive recommendation algorithms. TikTok uses machine learning to curate its "For You" feed.[11] This is known as being hyper-personalized and differs from other social media in that it relies on specific user input and engagement rather than traditional likes, comments, and follows. The company aims to be the top destination for short-form mobile video. TikTok's recommendation strategy uses a combination of computer vision, Natural Language Processing, and metadata tagging, quickly helping TikTok to become one of the most popular web domains in 2021. The company is still hiring machine-learning engineers, as it develops scalable classifiers, tools, models, and algorithms, for both data mining and computer vision.[12]

Swipe Left, Swipe Right

It turns out that finding a job and finding love are similar in a technological sense. A corollary use of matching technology is for dating apps.[13]

Dating systems use recommendation engines similar to movie and TV streaming platforms. Also, dating apps face the same problems as social media platforms in terms of hate speech, bullying, and bad actors, all of which can be helped by machine learning. Tinder formerly used an ELO rating system similar to chess rankings. The more people swiped right on your profile, the higher your ranking rose. (On many dating apps, swiping right on a picture of a prospective date means accept, and swiping left means reject.) Scores were boosted by the number of swipes and by the ELO scores of the swipers themselves. Tinder now says that technology is out of date, and such rankings have been replaced by more sophisticated algorithms. Tinder, similar to Facebook, uses machine learning to curb hate speech on the platform. OkCupid uses something called an interaction matrix, which takes users' past preferences to determine potential matches. Each user's decision of a like or dislike of a potential date is termed a vote. These votes are put into a matrix that can then interpolate future predictions. OkCupid uses techniques called "Gradient descent and singular value decomposition" to do this.[14] Bumble has expanded beyond dating to business opportunities and to find friendship. Bumble is designed so that women reach out first. Bumble created a multilingual toxicity detector that uses machine learning to protect users from harmful messaging. The feature works across 50 countries and 150 languages. The Bumble engineers used a three-class classification to scan for sexual, insults, and identity hate. They built a dataset of 3 million messages collected and labeled by hand over a period of months, with a series of manual checks for data quality. That system uses a transformer.[15] Hinge, meanwhile, uses all your cumulative data to find a match with its "Most Compatible" service. The algorithm uses the likes you send and receive over time to build up a profile of date that should prove mutually compatible. Hinge uses a Nobel Prize-winning matching algorithm from the 1960s called Gale–Shapley. The app doesn't require swiping, as with other dating apps. Instead, it aims to find matches that are longer-term in nature based on a wider array of data.[16] Chinese dating app Tantan used facial recognition to check that users aren't retouching their photos to deceive potential dates.[17]

Targeted Persuasion

Social media companies' use of data for persuasion and targeted advertising tactics is one area of worry for researchers. The dangers of abuse are very real, says computer scientist Yoshua Bengio, who highlights this as one of the bigger risks in artificial intelligence.[18] The use of these technologies in advertising and in persuading or influencing consumers is something that endangers democracy, he says. Also, as already mentioned, curbing hate speech and fake news across platforms using AI is harder than it initially appeared. Facebook struggled to stop Russian election meddling. LinkedIn failed to stop a hacker from stealing its data. AI is not a panacea for the problems of running social media platforms at scale.

New jobs and roles abound in social media platforms and dating apps. There also has been growing demand for skills in moderating content, although many platforms prefer to use technology over people, due to scale. Social media is now moving into live broadcasting and video as how customers use the platforms changes.

Chapter 23

Virtual Assistance

In the following pages, we assess the rise and plateau of voice assistants. Siri, Cortana, Google, and Alexa are getting better but are still imperfect helpers. We also identify how virtual assistants are improving their functionality as the underlying datasets increase in size. We look at the ways in which they will become more helpful, and some of the ways they won't, with an examination of their limitations and why they are still so amusingly imperfect. Eventually, as the technology evolves, virtual assistants are likely to be our digital future.

Virtual assistants have been around for a decade and more, but machine learning is only now starting to make them properly useful. And widespread global smartphone usage means they are ubiquitous and at the disposal of almost everyone. Many people don't know that machine learning is powering these systems or how it works. Because of improvements in natural language understanding and natural language processing, Siri and Alexa can now answer harder questions better every day. Virtual AI assistants use voice recognition to listen to questions or requests and then answer or complete tasks, like scheduling or operating things around the home. They can still make comical errors and it can be frustrating if your questions aren't understood by the machine. They're now available in smartphones, as well as in purpose-built devices such as Amazon's Alexa, which operates independently but connects to other digital items. These assistants combine microphones with speakers embedded into devices, and typically leverage machine learning from elsewhere. They can also control lights, heating or thermostats, door locks, TV, and music and respond to vocal commands, making them increasingly part of the wired home.

Compared with chatbots, virtual assistants add an extra language layer. If you are chatting online with a customer service bot, it's usually in text via instant message on a computer. When you ask a question of your phone, the machine needs to understand your voice, translate that voice to text, and then transform that text into numbers, raw data, before it can even start to hunt for answers. If it can't understand the question, the answer is doomed. Virtual assistants start with activation, via the press of a button or a "wake word" — a word that activates the system and tells it to anticipate a command or a query. "OK Google" or "Hey Siri" spring to mind. Next, there's an indicator that the assistant is ready for you. For Alexa, it's a blue light. For Siri, it's the Apple rainbow pattern, and for Google, it's circles in the primary colors. Amazon matches your acoustic patterns to the wake word to ensure you have hailed the machine, in a technique called keyword spotting. Next, Alexa accesses Amazon's cloud and is verified. There is some encryption, so the inquiry is kept secure. Then an answer is sent back. Alexa has microphones, and sometimes cameras, that can be controlled by the Echo device. Alexa also records your voice, and recordings can be managed on your account.[1] Amazon trains Alexa on your queries and commands so the system improves over time, and uses speech recognition based on natural language understanding. It's trained on real-world data from a wide range of customer speech patterns, accents, and dialects, as well as different acoustics. The training relies on supervised machine learning, where people label a small sample of requests with the correct responses.[2] The development system on which Alexa is based is called Amazon Lex and combines natural language understanding with Amazon's automatic speech recognition. That system is used for building and deploying interfaces involving conversations.[3] Amazon has also whitelisted the service to sell to third-party companies for use in their own chatbots.

OK Google! Hey Siri!

Google Assistant works in a similar way. But as with its search engine, Google works to figure out the intent behind your request.[4] It analyzes the text, along with recent searches, and looks at the type of device you're using. Google provides a list of different versions of what you might mean, along with the responses, and then ranks them. Google checks how sure it is that it's got your request correct and looks whether it has response. It examines how satisfied previous users were with similar requests. It also looks at how recent the response is, to ensure freshness.[5] Google provides developers with Dialogflow, a natural language understanding interface for conversation that can be used to build AI chatbots, and Actions on Google, a platform for developers to extend Google Assistant.[6]

Siri, meanwhile, activates with "Hey Siri" or a button press. Siri can answer questions, control apps on your iPhone, connect you with contacts, interact with text messaging, and control music.[7] A small speech recognizer runs constantly and just listens for the two wake-up words. It uses a deep neural network (DNN) to turn your voice's acoustic pattern into a probability confidence score of how likely it is that you spoke the words "Hey Siri," at which point Siri wakes up if that number is high enough. Apple says that Siri is the virtual assistant with the strongest privacy. Your voice input is processed by the phone and text transcripts of requests are sent to Apple, but they aren't associated with your Apple ID and the data are stored for a limited time. Also, your audio doesn't leave the phone, and Apple doesn't share your requests with advertisers.[8] Siri developers can use the system SiriKit.[9] Apple is working on Siri upgrades in the coming years to integrate the latest technologies and improve the results.

Microsoft shut down its virtual assistant Cortana on Android and iOS phones in 2021, after it fell behind the other voice helpers. Instead, the company took a different tack, making it available through Alexa and working as an app that can be used across various platforms rather than a Windows-exclusive product. Microsoft CEO Satya Nadella said in 2019 that he no longer views Siri or Alexa as competitors.[10] This shows the way that these technologies evolve and how users adopt them isn't always clear-cut, even for the biggest players. The company found building unique hardware and software for Cortana a challenge, so it went in a different direction. (Also, the public didn't love Cortana in the same way as Siri and Alexa, although people certainly have love–hate relationships with all these voicebots.) These all point to the hurdles of getting users to adopt complex products.

Smart Future

Gartner predicts 50% of knowledge workers will use a virtual assistant daily by 2025, from 2% in 2019.[11] Two problems exist for smart assistants that need to be circumvented for progress in this sphere: First, correctly understanding the voice commands and questions and second, handling the breadth of requests. Solving these two issues will be a tipping point for voice assistants both in usage and adoption.[12] They need to be powerful enough to process myriad data points, and they will need to help users choose between multiple different categories of decision, rather than just two. But once they manage this, smart assistants will be a convenient and powerful helper in your day-to-day life, embedded in every device around you: cars, mirrors, fridge, and phones. The future of virtual assistants is also likely to cross over between personal devices and work-based Robotic Process Automation office robots, which automate data and scheduling tasks. This can already be seen in how people are using ChatGPT and other large-language models. Also, virtual assistants' user interface of voice commands will spread to other applications and systems, as people get more acclimatized to interacting with machines verbally and as the speech recognition software improves. Separately, the future for virtual assistants may well be linked to the metaverse, depending on how that trend develops and which technologies get adopted. Devices such as the mouse or keyboard will be redundant in virtual reality, while voice is a natural interface, and will be crucial for making the metaverse a seamless social experience.[13] As the giant language models underpinning all these virtual assistants explode in size, as is currently happening, both performance and human-like abilities will dramatically improve in ways that will transform how we interact with computers and machines, hopefully for the better. It seems likely that more user interfaces will move toward vocal commands and away from text input over time, as they are so much easier to interact with. Picture a virtual assistant that has the narrative skills of ChatGPT, combined with the search power of Google and with the comprehension of Siri or Alexa on steroids. You now start to get something truly useful, at both answering questions and scheduling — if you hook it up to your calendar — checking facts, or providing serious help with your chores. This is likely to be the direction that virtual assistants will move in the future, but as we have seen, the ability to combine such tasks and cross-pollinate such technologies in this manner is no mean feat.

"They were all dumb as a rock," Microsoft Chief Executive Satya Nadella told the *Financial Times*. "Whether it's Cortana or Alexa or Google Assistant or Siri, all these just don't work. We had a product that was supposed to be the new front-end to a lot of [information] that didn't work."[14] While they have improved and continue to improve, there is a long way to go before we can rely on virtual helpers for the more complex requests.

Chapter 24

Access Denied

Lock the Front Vault! Machine learning is changing cybersecurity and access control systems. While it's helping sharpen security tools, hackers are hijacking the same technology, keeping security professionals on their toes. In this chapter, we explain why machine learning is key in this age of new online security technologies and why access control is such an essential element for critical infrastructure. Criminals are now using algorithmic techniques and machine learning too, making top-notch cybersecurity even more important. We examine the multiple benefits of access control, as systems become more flexible and safer. These systems are used to guard and safekeep the critical infrastructure of many countries, so the stakes are high, and sophisticated cyber-attacks are a growing risk.

When you go to work, you might badge in to log your identity and gain permission to enter your office. Often multiple badging-in is needed for the building entrance and then at your company or department. When you log into your computer network, your credentials ensure you have the correct access. Access control can limit physical access as well as virtual or online access and is key to both corporate security and governmental safety. Identity and access management systems are crucially important in a wired world. So is cyber security. It's natural that machine learning should be at the forefront of internet security because as the technologies change and adapt, the threats change too. Machine learning is perfect for sifting huge amounts of data to look for anomalies that might turn out to be threats. It can also save time and resources spent searching manually for such threats or hours spent controlling and verifying access and entry permissions. Access control is a fundamental part of security to determine who has the ability to access which systems.

Harnessed by Hackers

Unfortunately for the gatekeepers, the same technology can be harnessed by hackers and other bad actors to infiltrate systems, so it becomes a game of cat and mouse. Cybercriminals are able to use similar techniques to look for weaknesses. Artificial intelligence systems are also able to be tricked. Spam, phishing emails, and password guessing all risk getting supercharged by hackers using machine learning.[1] Deep fake tools are also a major risk to security, as they can be used to generate video or speech that's hard to tell from the real thing. A bank in Hong Kong was scammed into sending $35 million to criminals using a deep-fake voice recording that sounded like a real company executive.[2] So security experts need to stay abreast of how this technology is being used to neutralize existing tools, and access control is the first line of defense. You must control who has access to your systems and offices.

There are four broad kinds of access control: discretionary, mandatory, role-based, and attribute-based.[3] In discretionary control, the administrator sets the access policy. In mandatory control, there are set information clearance levels, similar to the military. For role-based access, business roles are defined with a specific level of access, rather than relying on identity. Attribute-based clearance is based on a set of attributes, such as the time of day, or location. The reason access control is so important is that it keeps data safe, preventing theft and protecting personal or sensitive information. As systems get more complex, more and more effort is needed to police access control systems to ensure the right people have the right access. This is where machine learning comes in, to save manpower and help monitor large systems. It also allows rigid control systems to be made more flexible.

In terms of broader security, machine learning can be used to find threats on a network, keep people safe while browsing the Web, protect against malware, and secure data in the cloud.[4] Data science models are adept at sifting through large datasets to detect anomalies. That lets systems check for network threats in real time, for example, from malware or attacks, policy violations, or even insiders. In browsing, they can help avoid "bad neighborhoods" online, stopping users from linking to malicious websites. Systems can be trained to detect and stop new malware, based on the behavior of known malware. Finally, they can assess suspicious cloud login activity and discrepancies in locations to identify threats in the cloud. A growing number of companies are harnessing machine learning to solve security problems.[5] That means there's a growing intersection of the two areas.

Money-Driven Hackers

Meantime, hackers are increasingly motivated by financial gains, making them more determined. Accenture Labs uses Nvidia GPUs and libraries to sniff out security threats quickly and at scale using large-scale network graphs.[6] It's a more proactive method, moving from a preventative approach toward hunting down attackers. Cyber security company Crowdstrike, which counts 61 of the Fortune 100 companies as customers, says machine learning is critical in defense against modern malware. Crowdstrike has observed an increase in malware. Attacks are increasingly no longer one-offs, meaning a shift toward continuous monitoring is required. The rise in remote work is also increasing cyber-security risks.[7] Artificial intelligence security firm Darktrace used its AI-centered defense in 2021 to take down a variant of TrickBot, which had become among the most prevalent malware in the world.[8] TrickBot was developed by sophisticated cybercriminals to steal financial data and then broadened to conduct an array of cybercrimes, according to the FBI.[9]

Patterns and Anomalies

Within security, machine learning usage largely falls under pattern recognition or anomaly detection. That's to say, uncovering hidden patterns either implicit or explicit from the data, or finding deviations from the norm that could be regarded as suspicious. Viruses, malware, worms, Trojan horses, and ransomware are all among the types of security threats. Also, safety and security engineers need to be aware that hackers are also using machine-learning techniques to boost their own efforts to penetrate online defenses, and so becoming more sophisticated. Cyber security beyond companies at the national level is also increasingly in the spotlight. A surge in state-sponsored attacks globally has prompted more focus on cybersecurity across countries.[10] In 2017, a cyber-attack called WannaCry hit more than 200,000 computers in 150 countries. It exploited a vulnerability found by the National Security Agency and then stolen and put online.[11] The ransomware attack targeted individuals, companies including FedEx, and government organizations like the British National Health Service. Total losses from the breach climbed as high as $4 billion. The US government said North Korea was to blame.[12]

Cybersecurity can become more effective and cheaper by harnessing machine learning techniques. Certain tasks can be helpful in discovering vulnerabilities and disrupting attacks. Yet further "significant breakthroughs" will be needed for a transformative impact of machine learning, according to research from Georgetown University in Washington DC.[13] That means incremental advances in cyber-defense are more likely than full-scale industry upheaval. Machine learning could also subtly alter the landscape for both attackers and defenders. Georgetown's report concludes that policymakers and practitioners should focus on specific tasks rather than talking in general terms of cybersecurity as a whole. That's likely good advice in other fields too.

Related to our analysis of autonomous weapons, one more existential risk identified by computer science professor Stuart Russell is the chance of cyber infiltration of such weapons platforms in the future.[14] That would mean that a defense system you think is under your control has been compromised without your knowledge and could turn on you when conflict starts. This introduces further strategic uncertainty in warfare. Another weak point for hackers is a country's critical infrastructure: The power grid, water supplies, telecommunications networks, factories, and ports. As these systems move more and more online and rely on ever more complex computer systems, they are at risk from hackers and cybercriminals. This applies to the US and elsewhere. Ukraine and Iran are recent examples where infrastructure has been targeted.

Cyber Field

Back to cybersecurity, the future looks sure to include machine learning. Certainly, a prerequisite for solid machine learning security is that the machine learning itself is robust and secure.[15] The future of access control is likely to be dominated by biometrics, as it's both more convenient for users and safer to rely on unique markers like fingerprints, iris scans, or facial recognition to unlock systems. Cybersecurity is a growing field in itself, and the job market is expanding as the needs spread across government and the corporate sector. Every cyber-threat or attack that emerges sends companies and governments scrambling for quick solutions. Also as the world becomes more wired, the risk of cyber-threats increases. This means the need for cross-over positions to handle security where traditional industries are moving online, as well as the increasingly sophisticated threats against technology companies, banks, and government infrastructure. Clearly, the risks of advances in cyber security come with concurrent risks of hackers exploiting them, so the cyber security industry is under pressure to stay one step ahead of criminals and bad actors.

Chapter 25

Wild at Heart

Capturing and analyzing data can aid wild animals. In this chapter, we focus on how the latest sensors and datasets can assist with large-scale data collection and analysis to help keep watch on endangered species. Some examples are detecting and counting animal populations, spotting poachers, measuring acoustics, sound-based surveillance, and checking on levels of plastic pollution. We go on to explain how data scientists can help wild elephants in their natural habitat and interpret whale song.

One of the unlikeliest, and most unusual, applications of machine learning is in wildlife conservation. It turns out that sensors are allowing data-gathering to monitor wildlife at levels unseen in the past. That means data science can be harnessed to monitor and protect wildlife at a scale impossible with human volunteers or professional conservationists. The technology is taking off throughout academia, often in partnership with industry. For example, the Center for Biodiversity and Conservation at the American Museum of Natural History has a project to promote machine learning in conservation. One effort, called the Animal Detection Network, automatically identifies and counts species of animals in images, audios, and videos:

The role of artificial intelligence in sustainability is not only nascent, but thriving among the many research groups, start-ups, established companies, development agencies and non-profit organizations which span the globe.[1]

Elephant Support

Another example of real-world impact is in helping to monitor and protect elephants. Elephants range far and wide across vast territories and are hard to track, and poachers pose a threat, so capturing their data is key to their protection. Even getting accurate elephant numbers is an important step to stopping population declines. African elephant numbers have dwindled in the past century because of poaching, habitat destruction, and retaliation for crop damage. Knowing where elephants are, and how many there are, is vital for various conservation groups in order to ensure their safety. But elephants aren't easy to monitor or tally. A team from the University of Oxford used Convolutional Neural Networks (CNNs) to detect and count elephants in South Africa from satellite images. The researchers estimated their model is just as accurate as humans at finding elephants in such images, and, it's possible, at a far greater scale.[2] This team made a training dataset of more than 1,000 elephants that was fed into a CNN. This technique could be broadened to identify other wild animals from satellites. Separately, the Save the Elephants project, a charity fighting poaching and the ivory trade, uses GPS technology and algorithms to monitor elephant location. They can also detect signs of stress in real time, such as more nighttime activity or streaking behavior, where elephants speed up along specific corridors.[3] The non-profit worked with WildMe, which has also created algorithms to recognize zebras and humpback whales.[4]

Hours of Whale Song

Ann Allen, an ecologist at the National Oceanic and Atmospheric Administration, has a model to identify humpback whale songs. With help from Google, she sifted through 180,000 hours of underwater recordings with 10 hours of annotated data identifying whale songs and other noises, to train a neural network to detect the songs. The results are helping her research on the whale species in the Pacific Ocean and how it has changed over the past decade.[5]

Other researchers have figured out how to use facial recognition to identify and track elephants and send alerts when poachers are nearby. Elephant facial recognition works similarly to facial recognition in humans, in that each elephant has a unique facial profile that can be codified as a number. This technology was built by The London Zoological Society and Google.[6] All these elephant conservation projects show the multitude of inventive ways machine learning can be used. The Elephant Listening Project uses infrasound to record elephant calls way below the level of human hearing, in an effort to study communication. Because there are about 1 million hours of tape, deep learning is one avenue to analyze these amounts of data. The Elephant Listening Project's 50 sensors are dotted around a 1,250 square kilometer area of a national park in the northern Republic of Congo. Paths are inaccessible, even by foot, but the acoustic grid picks up elephant call density, as well as gunshots. The project can quantify whether poaching is occurring and direct anti-poaching patrols. Researchers can use acoustics to map changes to the elephants' use of the landscape. There are two main challenges: getting the data out in usable form and pinpointing the useful sounds in the hours or months of tape.[7]

Cross-Pollination

More broadly, machine learning can, and should, help ecologists and wildlife conservationists make sense of the burgeoning streams of data to combat poaching and loss of diversity, as well as unravel some of the mysteries of animal behavior. An interdisciplinary community of computer scientists and ecologists is growing, which will be key to any success in the field, a study in *Nature* found. That's the familiar path followed elsewhere in industry, of combining domain expertise with specialist technical skills. Also, it will let local studies scale:

Animal ecology and wildlife conservation need to make sense of large and ever-increasing streams of data to provide accurate estimations of populations, understand animal behavior and fight against poaching and loss of biodiversity. Machine and deep learning bring the promise of being the right tools to scale local studies to a global understanding of the animal world.[8]

One issue with machine learning in conservation is that it can point out illegal behavior and highlight and pinpoint problems but not necessarily enforce solutions. Political will and resources are still needed to enact changes. So, this technology is a tool that can only go so far.

But in terms of watching habitats and ecosystems at scale, monitoring unsupervised, and sifting through the data automatically and highlighting areas of interest, machine learning has the potential to save researchers countless hours of time. The World Wildlife Fund says artificial intelligence is fundamentally altering the way we study and protect the natural world because it can predict the extinction of species, measure the impact of industry, and help stop poaching. It could also be used to monitor climate change and water quality.[9]

Plastic Peril

Plastic pollution is one of the world's most urgent environmental issues. The world is awash in waste plastic, endangering animals and humans in myriad ways. The UN Environment Program joined with Google and used geospatial data and annotated images to create a model through which to generate a detailed and accurate view of plastic waste in the Mekong River[10] in Southeast Asia. Scientists don't know enough about the leakage of plastic waste into rivers. This project looks to examine that lack of data. Australian non-profit Minderoo Foundation developed a tool that maps plastic pollution globally in real time.[11] Global Plastic Watch combines artificial intelligence with remote sensing satellite imaging to build an open-source high-resolution map.[12] Again, all these projects can and should be using this cutting-edge technology to help with their missions. They should not, however, be driven by technology or build systems for the sake of it. A root problem must exist for the technology to solve.

Using image recognition to detect different species is one way that Arkansas researchers looked at using feeders to dispense food to certain creatures and not others.[13] In one experiment, they used a device called a WiseEye smart feeder to see if it could differentiate between deer and raccoons. The ultimate aim was to control invasive feral pigs. The WiseEye feeder has a camera and a feedbox that only opens if the right species comes into view.[14] It also has an electric shock function to deter pests. Of course, building machine learning systems is difficult, expensive, and not necessarily suitable for wild habitats. Also, internet connections or even electricity may not be available in the various wildernesses where study is needed. Wildlife researchers and non-profits must rely on technology companies to provide support if they want to build large-scale systems. But as with the art world, usage of the technology is emerging in surprising and unforeseen ways. And companies such as Google are engaging with non-profits to help leverage their technology in the natural world, as it also improves their image and can be an easy public relations win. DeepMind built a model to monitor the behavior of animals in the Serengeti National Park in Tanzania.[15] Over nine years, millions of photos have been collected by ecologists, which until now needed to be checked by hand. The system will help detect, identify, and count the animals, helping unclog a bottleneck for the conservation researchers.

Wild Careers

Wildlife artificial intelligence researchers may seem like an unlikely career path, but naturalists with an understanding of big data will be more in demand as sensor systems proliferate in the non-profit world, and wildlife monitoring gets more acclimatized to using the latest data gathering and data analysis techniques. It's certainly a nascent industry but a growing one. There's also the worry about misuse of artificial intelligence in conservation, because of opacity, data scope, and bias, so greater care should be taken with understanding how the models work.[16] Machine learning is helping with predicting extinction risks of species, assessing the footprint of fisheries, and mapping and identifying animals in the field. Now there are growing calls for ethical considerations and more oversight to avoid unfair bias and misuse. The lower costs of sensors will have a key impact on the future of wildlife conservation. The potential for large-scale ecological studies is great but, at current, limited by the current approach to data processing. Distilling data into relevant information is inefficient.[17] The way forward is to combine approaches and integrate modeling tools and train a new generation of data scientists in conservation.

Chapter 26

Generative Generation

Generative AI systems are prompting excitement to the point of hysteria across industry, finance, and the media (if not academia). All this buzz about the latest chatbots and image generators is based on large language models. Is this the future of AI, and how will these developing trends change the workplace? In what follows, we provide an outlook on the prospects for the future of large language models, including some of their real-world uses, their rapid uptake, and their dangers.

Large-language models and the generative AI systems built with them provoke such a high level of interest because it really seems like they provide intelligent answers to questions and coherent responses to prompts. They can write or draw pictures, compile reports, write poetry, or hold a conversation, all at high speed, with minimal instructions. People can quickly see their possible uses across a wide array of tasks in the office or at home. The newest generation of chatbots particularly have wider scope than the kinds of narrow AI we have previously defined, even though they are still nowhere near Artificial General Intelligence or human-level understanding. As a reminder, large language models are neural networks with billions of weights and trained on large quantities of text using methods of unsupervised learning or self-supervised learning. LLMs can recognize, summarize, and translate. They can predict words

171

and generate text based on the mounds of data on which they are trained.[1] The occasional suspect accuracy of the answers, along with often dubious source material (often almost the entire text of the internet, for better or worse), is considered a price worth paying for such a useful tool with such cool applications. Also the interfaces are easy to use, making the appeal far wider than the original audience of technically savvy developers who were able to interact with earlier systems. The fact that you don't need to be able to code, you don't need to set up a complex application programming interface (API), or use an arcane query language to interrogate a knowledge graph database all lowers the barriers for entry to try out the latest version of these tools.

And the hype has exploded. The fact that machine-generated responses look so human makes people think that these systems really are Artificial General Intelligence, or even sentient, when this is, in fact, far from the case. ChatGPT was the fastest-adopted web platform in history.[2] The ease of use of the interface contributed to that. Google's Bard is similarly straightforward to use. Media coverage is increasing along with the usage. News stories and features related to artificial intelligence systems, particularly ChatGPT, have exploded in recent years. Google Trends analytics shows that news searches for the term AI climbed in 2022, jumping in November after ChatGPT's launch, and have continued rising since. Artificial Intelligence is experiencing a funding boom[3] from venture capitalists as other parts of the technology industry suffer retrenchment, job cuts, and a drying up of financing as global interest rates rise. A lot of the new funds are going toward generative AI and systems built on large language models, precisely because venture capitalists foresee real-world uses of the systems and real money to be made soon in the future.

Making Things Up

What AI developers gently describe as "hallucinations" have been common in ChatGPT and their ilk. In other words, the algorithm sometimes makes stuff up but does so in a very convincing way. Again, these machines have no concept of lies or truth, or right or wrong, but only data inputs and responses. But the application of a generative tool that tells lies or gets things wrong is by nature very different from one that always returns true facts and data. The machine often just provides answers that are plainly incorrect. As the internet contains hate speech and misinformation, and as large language models are trained on the internet, such chatbots are likely to be at risk of regurgitating or retooling hate speech and misinformation.[4] OpenAI has filters in place to prevent hate speech, but it's hard to prevent it completely when the very subject matter on which the model is trained contains it. Further, many users deliberately "jail break" the systems to get around such limitations and restraints, letting them generate any kind of offensive content that is requested.

Safety Harness

How are businesses harnessing these technologies when they are in search at an early stage and with so many seeming risks? US investment bank Morgan Stanley adopted OpenAI's GPT-4 system to organize its wealth management information. The bank has a library of data and analysis on insights for investment strategies, market research, commentary, and analyst insights. These are housed across internal sites, mostly as portable document format (PDF) files, historically a tricky way for machines to extract information. That meant employees needed to scan a vast array of information to answer specific questions, a laborious process that took time and effort. Morgan Stanley in 2022 started using GPT-3 and GPT-4, specifically the embeddings and retrieval technology, to build an internal chatbot to search across its wealth management content. This interface parses the data in a way that is both usable and actionable, according to the bank. "It effectively unlocks the cumulative knowledge of Morgan Stanley Wealth Management," says Jeff McMillan, head of analytics and data and innovation at the unit.[5]

Global accounting and consulting firm PWC partnered with AI startup Harvey for a chatbot built on GPT-4 for its 4,000 lawyers and legal staff to use. Harvey is a startup backed by OpenAI, focused specifically on lawyers. PWC promised that any machine-generated output used will be reviewed by its staff. The firm said it will customize the models for its own legal services business. It will also work to make the platform available to clients to streamline their own internal legal processes.[6] Payments company Stripe uses GPT-4 to scan the web to understand its users' businesses. It also uses the technology in customer support for questions around documentation.[7] It's not just companies — governments are examining the technology too. The government of Iceland is using GPT-4 to help preserve its own language. Because large language models don't have much training in Icelandic, they can't work in the same way as language translation services in more popular languages. Iceland wanted to change this by helping train models using reinforcement learning from 40 human volunteers to improve Icelandic culture and language.[8] We have already referenced Duolingo, but that language education company is also harnessing GPT-4 for its users trying to learn foreign languages to help explain their mistakes, along with an added feature that lets them ask questions, and the model can personalize feedback to students. Duolingo Max lets users ask why an answer was right or wrong and get an explanation that helps instruct their learning. The company's Roleplay function allows students to practice real-world conversations to help their language abilities.[9]

Beyond chatbots, what are the advantages of the latest iterations of these large language models? One key innovation is the ability of the machine to write computer code for you. Developers can give prompts or more detailed instructions, and the system will write them an entire program in a computer language of their choosing. This is one application that will be transformative for the software world.[10] OpenAI built a system called Codex to write code automatically, translating human language to over a dozen computer programming languages.[11] That system ceased and was folded into OpenAI's other API interfaces in 2023. Similar code-writing systems exist at DeepMind and elsewhere and are getting a lot of attention in the software community, some of whom are worried about the longevity of their own jobs. Some language models can interpret images as well as text. This is a significant step beyond a narrow machine learning application toward broader usage.[12] If machines can cross these boundaries and describe images using text, or draw based on instructions in words, then potential uses expand and the technology advances in leaps and bounds.[13] One point to note is that generative chatbots are not the same as search engines. In fact, they are totally different in both design and aim. While search engines endeavor to find facts or answers to questions on the internet and return results that match the user's search, generative bots are just predicting the next word in a sentence, and the word after that, and the word after that. This means that the intention of generative AI isn't to provide true answers or search and retrieve information but only to provide linguistic clarity based on statistics. Also, language models are currently based on some set of historical training data, meaning that they don't operate in real time, another facet that might trip users up if they are hoping for real-time news or answers to current questions.

Emergent Rules

As the models and computing power expand, so will the input size or the amount of words the systems will be able to handle. Users will be able to paste the text of whole books rather than just a few sentences. Machine-generated responses will get smarter and faster. With this comes risks of students cheating at tests, whole jobs being made superfluous as they can now be achieved with one mouse click, and the creation of ever more realistic fakes, cleverer phishing emails, and other malware. Italy was the first Western country to ban ChatGPT, in early 2023, deciding that data collection methods were too intrusive and breached its data privacy rules.[14,15] Other countries are fast examining making rules for AI and for large language models and their use. Britain is examining its own AI regulations, and the European Union is also looking at how to police the technology. The White House released what it called a blueprint for an AI Bill of Rights in 2022. Japan has plans to lead formulating international rules on AI.[16] And China plans administrative measures for generative Artificial Intelligence. As the technology shifts, rules will be inevitable. Governments need to balance the fine line between stifling innovation and keeping their citizens safe from online harm. Either way, regulation is coming.

Conclusion

DeepMind's project to predict the shape of proteins in three dimensions is considered to be one of the biggest achievements in artificial intelligence ever. The release of AlphaFold in 2020,[1] and the subsequent plan to update the database to include predictions of more than 200 million proteins, not only breaks scientific ground and solves a 50-year problem, but it also provides scientists with a fertile area to explore for years to come. Scientists say the breakthrough stands to transform biology.[2] This is an instance of where learning models were used as a tool to magnify both the speed and scale of a scientific project. Machine learning is an amazing technology with some surprising limitations. It's become embedded in our lives and will become further enmeshed in our world as people adopt new gadgets that transform how we interact with computers and communication devices and as companies push boundaries on exactly how we do those things. Hopefully, this book has helped unlock some of machine learning's secrets and shed light on how it works, how it's being used, and how it will be used in future.

We've examined ethics and education, and the risks and opportunities there, along with computer vision, text analysis, and machine translation. We've also looked at specific uses, such as the recommendation engines used in entertainment and retail, warehouse logistics and automation, and jobs and the economy. The main ethical concerns in both academia and industry are the transparency of algorithms, potential bias, risks of job losses, and economic upheaval through automation.

Scholars also worry about the loss of privacy and the risks surrounding data security. Specifically, facial recognition and autonomous weapons (with no humans in the loop) raise particular worries for practitioners. These are technologies that are available now, rather than far off in the future, and are relatively low cost compared with other weapons of mass destruction. A few academics worry about the rise of super-intelligence, where machines will overtake humans in what is known as the Singularity. Once machines become more intelligent than humans, they might be inclined to wipe out humanity, posing an existential risk for us. Many experts in computer science dismiss this over more prosaic but immediate concerns, but say the topic is still worth thinking about and discussing. The fact that some people are looking at such black swan events, considering their ramifications, and raising awareness can't be a bad thing. Some experts are changing their minds about machine learning and its ramification as the technology transforms and the breakthroughs accelerate. Geoffrey Hinton, the professor and practitioner who is one of the industry's main architects, had such a change of heart in 2023 when he quit Google and told *The New York Times* that part of him regrets his life's work. "It is hard to see how you can prevent the bad actors from using it for bad things," Hinton said.[3] He worries about a flood of fake videos, photos, and texts, along with the broad effects on the job market in the near term. Further out, he's concerned about possible unexpected behaviors and that machines will get smarter than humans faster than he earlier believed.

The benefits of machine learning include the ability to analyze this sea of data around us, predict diseases earlier, and scan information at scale. We can better detect money laundering and eventually will be able to reduce deaths on the road through self-driving cars. Hype in the industry can be avoided by understanding exactly what these specific applications can do and knowing their limits. Machine-learning systems are only as good as the data on which they rely and the designs of the engineers who built them. They also tend to be narrow rather than broad in their applications, meaning that a robot that can come over and make coffee in your home is still a distant proposition. Machine learning and data science are being used in new and surprising ways, such as in wildlife conservation, to track and protect wild elephants, and in art, music, and poetry. Artists are expanding the scope of how artificial intelligence can and should be used, as well as questioning some of its founding assumptions and exploring its darker sides. The world of work will be changed forever by these technologies as they quickly become production quality, reducing the need for clerical work and automating many rote and not-so-rote tasks around the office. There's opportunity as well as gloom here, as machine learning will create many new jobs and roles that will be needed to build, implement, and monitor different systems across multiple sectors of the economy. (A whole job sector is springing up around prompt engineers for generative AI, that can write useful commands and queries for chatbots.) Also, automation is mostly realized by replicating specific tasks rather than replacing whole jobs, so workers can, and must, navigate how this future will play out in order to adapt and keep their skills indispensable to the workplace. The technology giants, notably Google, Amazon, and Microsoft, are at the forefront of machine learning and artificial intelligence for now and have both embedded it in their core businesses and expanded their reach beyond their main businesses. They have the data to make deep learning fruitful. The rise of China in artificial intelligence is a worry for governments, but academics tend to see it as less of an arms race than policy makers do.

OpenAI's fast rise shows that you don't necessarily need to be a behemoth to produce cutting-edge tools that can find uses, but you do need a behemoth's level of data to make the system any good. Because of this and rival large language models, chatbots and virtual assistants are about to get a lot more powerful and hopefully more useful. Hedge funds were making money using artificial intelligence two decades ago and are still using it to great effect, getting an early heads up on the economic developments that affect market movements. Parsing the firehose of data into useful insights is the key to their success. Often, some of the more mundane uses of machine learning turn into some of the most useful cases for business. (Construction of asset portfolios comes to mind.) It's also becoming increasingly normalized and embedded in everyday gadgets. We routinely talk with customer service chatbots, ask our phones for guidance to a destination, or turn on the lights in our home with a voice command, all powered under the hood by learning models of some sort. It's going to get trickier to gauge whether there's a machine-learning component embedded in your application as this technology spreads everywhere — one key reason it's important to learn about is that it's controlling more and more of our lives. It's hard to know exactly what the future holds, but some of the trends we've seen are unstoppable, or at least genie-like, in that you can't put them back into the bottle once they're created. Maybe this or that company will win out, with this or that system, but the innovation in these technologies from academia to industry is an unrelenting force. It just needs to move (or be guided) in the right direction. The ladder needs to be placed against the right wall.

DeepMind's AlphaFold example of where technology can really scale and bring about profound change in the world is the shape of things to come in machine learning, as computer scientists work with domain experts on big problems that matter. Machine-learning algorithms should be powering advances in health care, finance, and science. They should be solving the hardest problems of economics and society, not just making clever chatbots or boosting advertising on social media platforms. Artificial intelligence should be harnessed to address humanity's biggest challenges, those of the environment, of underdeveloped economies, of changing workplaces; it shouldn't be constrained to small problems. But the industry is still at an early stage of the technological revolution. Watch this space.

Further Reading

Bengio, Y., et al. *Deep Learning*. UK, MIT Press, 2016.

Bostrom, Nick. *Superintelligence: Paths, Dangers, Strategies*. UK, Oxford University Press, 2014.

Broussard, Meredith. *Artificial Unintelligence: How Computers Misunderstand the World*. UK, MIT Press, 2019.

Chen, Qiufan and Lee, Kai-Fu. *AI 2041: Ten Visions for Our Future*. USA, Currency, 2021.

Coeckelbergh, Mark. *AI Ethics*. UK, MIT Press, 2020.

Domingos, Pedro. *The Master Algorithm: How the Quest for the Ultimate Learning Machine Will Remake Our World*. USA, Basic Books, 2015.

Ford, Martin. *Architects of Intelligence: The Truth about AI from the People Building It*. UK, Packt Publishing Limited, 2018.

Hinton, Geoffrey and Sejnowski, Terrence J. (eds.). *Unsupervised Learning: Foundations of Neural Computation*. UK, Bradford University Press, 1999.

Iansiti, Marco and Lakhani, Karim R. *Competing in the Age of AI: Strategy and Leadership When Algorithms and Networks Run the World*. USA, Harvard Business Review Press.

Marcus, Gary and Davis, Ernest. *Rebooting AI: Building Artificial Intelligence We Can Trust*. USA, Knopf Doubleday Publishing Group, 2020.

Martin, James H. and Jurafsky, Dan. *Speech and Language Processing: An Introduction to Natural Language Processing, Computational Linguistics, and Speech Recognition*. India, Prentice Hall, 2000.

Mason, Matthew T. *Mechanics of Robotic Manipulation*. MIT Press, 2001.

McAfee, A., *et al*. *Artificial Intelligence: The Insights You Need from Harvard Business Review*. UK, Harvard Business Review Press.

Metz, Cade. *Genius Makers: The Mavericks Who Brought AI to Google, Facebook, and the World*. New York, Dutton, an imprint of Penguin Random House LLc, 2021.

Murphy, Kevin P. *Machine Learning: A Probabilistic Perspective*. UK, MIT Press, 2012.

O'Neil, Cathy. *Weapons of Math Destruction: How Big Data Increases Inequality and Threatens Democracy*. UK, Crown, 2016.

Russell, S., *et al*. *Artificial Intelligence: A Modern Approach*. UK, Prentice Hall, 2009.

Sejnowski, Terrence J. *The Deep Learning Revolution*. MIT Press, 2018.

Vargas, Patricia A. and Aylett, Ruth. *Living with Robots: What Every Anxious Human Needs to Know*. USA, MIT Press, 2021.

Notes and References

All vectors used in this book are sourced from Shutterstock. "Simple Set of Robots Related Vector Line Icons," by davooda. Stock Vector ID: 488397142.

Introduction

1. Ng, A. How to Choose Your First AI Project. *Harvard Business Review*, February 2019. https://hbr.org/2019/02/how-to-choose-your-first-ai-project.
2. Rieley, M. Bureau of Labor Statistics, June 2018. https://www.bls.gov/opub/btn/volume-7/big-data-adds-up.htm.
3. Conn, A. Benefits and Risks of AI. Future of Life, November 2015. https://futureoflife.org/background/benefits-risks-of-artificial-intelligence/.
4. Bostrom, N. Ethical Issues in Advanced Artificial Intelligence. 2003. https://nickbostrom.com/ethics/ai.
5. Schwab, K. *The Fourth Industrial Revolution*. Portfolio Penguin, 2017.
6. Jurgens, J. World Economic Forum. Shaping the Future of Technology Governance: Artificial Intelligence and Machine Learning. https://www.weforum.org/publications/annual-report-2021-2022/in-full/centre-for-the-fourth-industrial-revolution/.

Chapter 1

1. Langston, J. Q&A with Pedro Domingos: Author of 'The Master Algorithm.' *University of Washington News,* September 17, 2015. https://www.washington.edu/news/2015/09/17/a-q-a-with-pedro-domingos-author-of-the-master-algorithm/.
2. IBM. What is machine learning? July 2020. https://www.ibm.com/cloud/learn/machine-learning.
3. Berkeley. What Is Machine Learning (ML)? June 26, 2020. https://ischoolonline.berkeley.edu/blog/what-is-machine-learning/.
4. Broussard, M. *Artificial Unintelligence: How Computers Misunderstand the World.* MIT Press, 2018.
5. Pretz, K. Stop Calling Everything AI, Machine-Learning Pioneer Says. IEEE Spectrum, March 2021. https://spectrum.ieee.org/stop-calling-everything-ai-machinelearning-pioneer-says.
6. Wozniak, S. Interview with Fast Company. *YouTube.*
7. Was, R. *AI and Machine Learning.* New Delhi, Sage Publications, 2020.
8. McCarthy, J. What is AI? http://jmc.stanford.edu/artificial-intelligence/what-is-ai/index.html.
9. Hinton, Geoffrey, cited in Ford, Martin. *Architects of Intelligence: The Truth about AI from the People Building It.* UK, Packt Publishing, 2018.
10. Samuel, A.L. Some studies in machine learning using the game of checkers, in IBM Journal of Research and Development, vol. 3, no. 3, pp. 210–229, July 1959. https://ieeexplore.ieee.org/document/5392560.
11. Kaplan, J. *Artificial Intelligence: What Everyone Needs to Know.* UK, Oxford University Press, 2016.

Chapter 2

1. Netflix. How Netflix's Recommendation System Works. https://help.netflix.com/en/node/100639?ba=SwiftypeResultClick&q=How%20Netflix%E2%80%99s%20Recommendations%20System%20Works.

2. Schrage, M. Great Digital Companies Build Great Recommendation Engines. *Harvard Business Review*, 2017. https://hbr.org/2017/08/great-digital-companies-build-great-recommendation-engines.

3. Stoll, J. Quarterly Netflix Subscribers Count Worldwide 2013-2023. 2023. https://www.statista.com/statistics/250934/quarterly-number-of-netflix-streaming-subscribers-worldwide/.

4. Yellin, T. Future of Storytelling Podcast, Episode 44, 2021. https://futureofstorytelling.org/story/todd-yellin-ep-44.

5. Madrigal, A. How Netflix Revers Engineered Hollywood. 2014. https://www.theatlantic.com/technology/archive/2014/01/how-netflix-reverse-engineered-hollywood/282679/.

6. Yellin, T. *op. cit.*

7. Netflix Codes. https://www.netflix-codes.com/.

8. What's on Netflix. https://www.whats-on-netflix.com/library/categories/.

9. Netflix. Tech Blog. 2012. http://techblog.netflix.com/2012/06/netflix-recommendations-beyond-5-stars.html.

10. Netflix. Recommendations. https://research.netflix.com/research-area/recommendations.

11. Hardesty, L. The History of Amazon's Recommendation Algorithm. 2019. https://www.amazon.science/the-history-of-amazons-recommendation-algorithm.

12. Smith and Linden. Two Decades of Recommender Systems at Amazon.com. 2017. https://assets.amazon.science/76/9e/7eac89c14a838746e91dde0a5e9f/two-decades-of-recommender-systems-at-amazon.pdf.

13. Hardesty, L. (2019), *op. cit.*

14. Chen, A. The Cold Start Problem: How to Start and Scale Network Effects. First ed. Harper Business an Imprint of HarperCollinsPublishers, 2021.

15. Schrage, (2017), *op. cit.*

16. *Ibid.*

Chapter 3

1. Google. How Search Works. https://www.google.com/search/howsearchworks/how-search-works/ranking-results/.

2. Rowe, K. How Search Engines Use Machine Learning: 9 Things We Know For Sure. *Search Engine Journal*, 2021. https://www.search enginejournal.com/ml-things-we-know/408882/#close.

3. Pichai, S. How I Built This with Guy Raz Podcast. https://podcasts. apple.com/us/podcast/hibt-lab-google-sundar-pichai/id1150510 297?i=1000564898021.

4. Levy, S. How Google is Remaking Itself as a "Machine Learning First" Company. *Wired Magazine*, 2016. https://www.wired.com/2016/06/ how-google-is-remaking-itself-as-a-machine-learning-first-company/.

5. Sullivan, D. How Google Autocomplete Predictions Work. Google. https://blog.google/products/search/how-google-autocomplete-predictions-work/.

6. Collins, E. and Ghahramani, Z. LaMDA: Our breakthrough conversation technology. Google Blog. https://blog.google/technology/ai/ lamda/.

7. Love, J. Google CEO Says Its ChatGPT Rival Coming Soon as a 'Companion' to Search. *Bloomberg*. https://www.bloomberg.com/news/ articles/2023-02-02/google-to-make-ai-language-models-available-soon-pichai-says?srnd=technology-vp&leadSource=uverify%20wall.

8. Goodrow, C. On YouTube's Recommendation System. YouTube Official Blog. YouTube. https://blog.youtube/inside-youtube/on-youtubes-recommendation-system/.

9. Microsoft Bing Blogs. AI at Scale in Bing. 2020. https://blogs.bing. com/search/2020_05/AI-at-Scale-in-Bing.

10. Microsoft Bing Blogs. Towards More Intelligent Search: Deep Learning for Query Semantics. 2018. https://blogs.bing.com/search-quality-insights/May-2018/Towards-More-Intelligent-Search-Deep-Learning-for-Query-Semantics.

11. *Ibid.*

12. Microsoft Bing Blogs, (2020), *op. cit.*

13. Nayak, P. Understanding Searches Better Than Ever Before. Google, 2019. https://blog.google/products/search/search-language-understanding-bert/.

14. Teofili, T. *Deep Learning for Search*. United States, Manning, 2019.

15. *Ibid.*

16. Halladay, K. Companies Are Desperate for Machine Learning Engineers. *BuiltinNYC*, 2022. https://builtin.com/data-science/demand-for-machine-learning-engineers.

Chapter 4

1. IBM. What is Computer Vision? https://www.ibm.com/topics/computer-vision#:~:text=Computer%20vision%20is%20a%20field, recommendations%20based%20on%20that%20information.
2. Markoff, J. How Many Computers to Identify a Cat? 16000. *New York Times*, 2012. https://www.nytimes.com/2012/06/26/technology/in-a-big-network-of-computers-evidence-of-machine-learning.html.
3. Dean, J., and Ng, A. Using Large-Scale Brain Simulations for Machine Learning and AI. Google Blog, 2012. https://blog.google/technology/ai/using-large-scale-brain-simulations-for/.
4. Luis Von Ahn, Interview With NPR. 2019. https://www.npr.org/transcripts/716827880.
5. Brownlee, J. Machine Learning Mastery. How Do Convolutional Layers Work in Deep Learning Neural Networks? Machine Learning Mastery, 2019. https://machinelearningmastery.com/convolutional-layers-for-deep-learning-neural-networks/.
6. Saha, S. A Comprehensive Guide to Convolutional Neural Networks — The ELI5 Way. Towards Data Science, 2018. https://towardsdatascience.com/a-comprehensive-guide-to-convolutional-neural-networks-the-eli5-way-3bd2b1164a53.
7. Pesenti, J. An Update On Our Use of Face Recognition. Facebook, 2021. https://about.fb.com/news/2021/11/update-on-use-of-face-recognition/.

Chapter 5

1. Stouffer, C. How Facial Recognition Works. Norton, 2023. https://us.norton.com/internetsecurity-iot-how-facial-recognition-software-works.html.

2. Madiega, T., and Mildebrath, H. European Parliament Document. Regulating Facial Recognition in the EU. https://www.europarl.europa.eu/RegData/etudes/IDAN/2021/698021/EPRS_IDA(2021)698021_EN.pdf.
3. Mohanakrishnan, R. Top 11 Facial Recognition Software in 2021. Spiceworks, 2021. https://www.spiceworks.com/it-security/identityaccess-management/articles/facial-recognition-software/.
4. Amazon. Detecting and Analyzing Faces. https://docs.aws.amazon.com/rekognition/latest/dg/faces.html.
5. Marr, B. *Artificial Intelligence in Practice*. Chichester, Wiley, 2019, p. 51.
6. Facebook Help Page. https://www.facebook.com/help/122175507864081.
7. Stouffer, C., *op. cit.*
8. Big Brother Watch. https://bigbrotherwatch.org.uk/campaigns/stop-facial-recognition/#facts.
9. Bird, S. Responsible AI Investments and Safeguards for Facial Recognition. Microsoft, 2022. https://azure.microsoft.com/en-us/blog/responsible-ai-investments-and-safeguards-for-facial-recognition/.
10. Smith, B. Facial Recognition: It's Time for Action. Microsoft, 2018. https://blogs.microsoft.com/on-the-issues/2018/12/06/facial-recognition-its-time-for-action/.
11. Microsoft. Principles on Facial Recognition. https://blogs.microsoft.com/wp-content/uploads/prod/sites/5/2018/12/MSFT-Principles-on-Facial-Recognition.pdf.
12. Crampton, N. Framework for AI Systems. Microsoft, 2022. https://blogs.microsoft.com/on-the-issues/2022/06/21/microsofts-framework-for-building-ai-systems-responsibly/.
13. Pesenti, J. Update on Use of Face Recognition. Facebook, 2021. https://about.fb.com/news/2021/11/update-on-use-of-face-recognition/.
14. Harwell, D. Amazon Facial Recognition Ban, *Washington Post*, 2021. https://www.washingtonpost.com/technology/2021/05/18/amazon-facial-recognition-ban/.
15. Electronic Frontier Foundation. Face Recognition. https://www.eff.org/pages/face-recognition.

16. Davies, D. Facial Recognition and Beyond: Journalist Ventures Inside China's 'Surveillance State.' NPR, 2021. https://www.npr.org/ 2021/01/05/953515627/facial-recognition-and-beyond-journalist- ventures-inside-chinas-surveillance-sta.
17. People's Daily Tweet. 2018. https://twitter.com/PDChina/status/ 978444380066390016.
18. CBInsights. Rise Of China's Big Tech In AI: What Baidu, Alibaba, and Tencent are Working On. 2018. https://www.cbinsights.com/research/ china-baidu-alibaba-tencent-artificial-intelligence-dominance/.
19. CBInsights. China Surveillance AI. 2018. https://www.cbinsights. com/research/china-surveillance-ai/.
20. Duo, E. China built the world's largest facial recognition system. 2021. *Washington Post*, https://www.washingtonpost.com/world/ facial-recognition-china-tech-data/2021/07/30/404c2e96-f049-11eb- 81b2-9b7061a582d8_story.html.
21. *Bloomberg*. China Fines Didi $1.2 Billion after Wrapping Year-Long Probe. 2022. https://www.bloomberg.com/news/articles/2022-07- 21/china-fines-didi-1-2-billion-after-wrapping-cybersecurity- probe?sref=S1DrA7g6.
22. Gargaro, D. The Pros and Cons of Facial Recognition Technology. IT Pro, 2022. https://www.itpro.com/security/privacy/356882/the- pros-and-cons-of-facial-recognition-technology.
23. Berle, I. *Face Recognition Technology: Compulsory Visibility and Its Impact on Privacy and the Confidentiality of Personal Identifiable Images*. Germany, Springer International Publishing, 2020.
24. Emergen Research. Top 10 leading Facial Recognition Companies in the World. April 2022. https://www.emergenresearch.com/blog/ top-10-leading-facial-recognition-companies-in-the-world.

Chapter 6

1. Marcus, G., Davis, E. *Rebooting AI: Building Artificial Intelligence We Can Trust*. 1st edn., Pantheon Books, 2019.

2. Taft, I. Police Use of Robot to Kill Dallas Suspect Unprecedented, Experts Say. Texas Tribune, 2016. https://www.texastribune.org/2016/07/08/use-robot-kill-dallas-suspect-first-experts-say/.

3. Møller, M. Secretary General Statement. United Nations, 2019. https://www.un.org/sg/en/content/sg/statement/2019-03-25/secretary-generals-message-meeting-of-the-group-of-governmental-experts-emerging-technologies-the-area-of-lethal-autonomous-weapons-systems.

4. Debusmann, B. Killer Robots and the Future of War. Vox, 2023. https://www.wionews.com/opinions-blogs/killer-robots-and-the-future-of-war-549249.

5. The UN is urgently calling for new rules as of October 2023. https://news.un.org/en/story/2023/10/1141922.

6. Hagstrom, M. Military Applications of Machine Learning and Autonomous Systems. The Impact of Artificial Intelligence on Strategic Stability and Nuclear Risk: Volume I, *Euro-Atlantic Perspectives*, edited by Vincent Boulain, Stockholm International Peace Research Institute, 2019, pp. 32–38. JSTOR. http://www.jstor.org/stable/resrep24525.10.

7. *Ibid.*

8. Department of Defense. Task Force Report, 2012. https://irp.fas.org/agency/dod/dsb/autonomy.pdf.

9. Etzioni, A., and Etzioni, O. Pros and Cons of Autonomous Weapons Systems. *Military Review*, 2017. https://www.armyupress.army.mil/Portals/7/military-review/Archives/English/pros-and-cons-of-autonomous-weapons-systems.pdf.

10. Department of Defense. *Ibid.*

11. UN Security Council Report. March 2021. https://documents-dds-ny.un.org/doc/UNDOC/GEN/N21/037/72/PDF/N2103772.pdf?OpenElement.

12. Russell, A., *et al.* Lethal Autonomous Weapons Exist; They Must Be Banned. 2021. https://spectrum.ieee.org/lethal-autonomous-weapons-exist-they-must-be-banned.

13. Aylett, R., and Vargas, P. *Living With Robots*. MIT Press. 2021. p. 224.

14. Peck, M. 1945. The US Military's Future: Waves of Killer Drone Swarms. December 2021. https://www.19fortyfive.com/2021/12/the-us-militarys-future-waves-of-killer-drone-swarms/.

15. UK Government Policy Paper, June 2022. https://www.gov.uk/government/publications/defence-artificial-intelligence-strategy/defence-artificial-intelligence-strategy.
16. Future of Life Institute. What are Autonomous Weapons? https://autonomousweapons.org/.
17. Huelss, H., and Bode, I. *Autonomous Weapons Systems and International Norms*. UK, McGill-Queen's University Press, 2022.
18. Lee, Kai-Fu. The Third Revolution in Warfare. The Atlantic. September, 2021. https://www.theatlantic.com/technology/archive/2021/09/i-weapons-are-third-revolution-warfare/620013/.

Chapter 7

1. Marcus, G. and Davis, E. *Rebooting AI: Building Artificial Intelligence We Can Trust*. First ed. Pantheon Books 2019.
2. Bass, D. Microsoft Invests $10 Billion in ChatGPT Maker OpenAI. *Bloomberg News*, 2023. https://www.bloomberg.com/news/articles/2023-01-23/microsoft-makes-multibillion-dollar-investment-in-openai?sref=S1DrA7g6.
3. Bloomberg Website. Bloomberg AI Researchers Advance Paraphrase Generation & Chart Summarization in AAAI 2022 Papers. February 2022. https://www.bloomberg.com/company/stories/bloomberg-ai-researchers-advance-paraphrase-generation-chart-summarization-aaai2022/.
4. TutorialsPoint. Natural Language Toolkit — Introduction. https://www.tutorialspoint.com/natural_language_toolkit/index.htm.
5. Davis, A.E. The Future of Law Firms (and Lawyers) in the Age of Artificial Intelligence. American Bar Association, 2020. https://www.americanbar.org/groups/professional_responsibility/publications/professional_lawyer/27/1/the-future-law-firms-and-lawyers-the-age-artificial-intelligence/#ref3.
6. Moulinier, I. How AI and Machine Learning is Shaping Legal Strategy. Thomson Reuters. https://www.thomsonreuters.com/en/careers/careers-blog/how-ai-and-machine-learning-is-shaping-legal-strategy.html.

Chapter 8

1. Ormond, J. Fathers of the Deep Learning Revolution Receive ACM A.M. Turing Award. ACM, 2019. https://www.acm.org/media-center/2019/march/turing-award-2018.
2. Hinton, G., *et al.* A Fast Learning Algorithm for Deep Belief Nets. 2006. https://www.cs.toronto.edu/~hinton/absps/fastnc.pdf.
3. Estela, A. The ALPAC Report. Pangeanic, 2013. https://blog.pangeanic.com/alpac-report#:~:text=The%20best%2Dknown%20event%20in,States%20for%20some%20twenty%20years.
4. Google Brain. https://research.google/teams/brain/.
5. Lewis-Kraus, G. The Great AI Awakening, *New York Times*, December 2016. https://www.nytimes.com/2016/12/14/magazine/the-great-ai-awakening.html.
6. *New York Times.*
7. Le, Quoc V., and Schuster, M. A Neural Network for Machine Translation, at Production Scale. Google AI Blog, 2016. https://ai.googleblog.com/2016/09/a-neural-network-for-machine.html.
8. Wu, Y., *et al.* Google's Neural Machine Translation System: Bridging the Gap between Human and Machine Translation. Google Research Paper, 2016. https://research.google/pubs/pub45610/.
9. Schuster, M., Johnson, M., and Thorat, N. Zero-Shot Translation with Google's Multilingual Neural Machine Translation System. Google AI Blog, 2016. https://ai.googleblog.com/2016/11/zero-shot-translation-with-googles.html.
10. Sun, M. *et al.* Baidu Neural Machine Translation Systems for WMT19. Research Papers on Machine Translation. Baidu, 2019. https://aclanthology.org/W19-5341/.
11. Etherington, D. Nvidia Breaks Records in Training AI. TechCrunch, 2019. https://techcrunch.com/2019/08/13/nvidia-breaks-records-in-training-and-inference-for-real-time-conversational-ai/.
12. Alarcon, N. Nvidia Developer Blog, 2019. https://developer.nvidia.com/blog/top-5-ai-speech-applications-using-nvidias-gpus-for-inference/.
13. Google. https://translate.google.com/intl/en/about/languages/.

14. Turovsky, B. Ten Years of Google Translate. Google, 2016. https://
blog.google/products/translate/ten-years-of-google-translate/
#:~:text=2.,3.
15. IBM. Speech Recognition. https://www.ibm.com/cloud/learn/speech-
recognition.
16. Lutkevich, B. Speech Recognition. Techtarget. https://www.techtar-
get.com/searchcustomerexperience/definition/speech-recognition.
17. Duolingo. Approach. https://www.duolingo.com/approach.
18. Peranandam, C. AI Helps Duolingo Personalize Language Learning.
Wired, 2018. https://www.wired.com/brandlab/2018/12/ai-helps-
duolingo-personalize-language-learning/#:~:text=To%20enable%20
this%20AI%2C%20Duolingo%20uses%20deep%20learning%
2C,to%20quickly%20analyze%20data%20and%20make%20
intelligent%20predictions.
19. Google Developer Site. https://developers.google.com/recaptcha/
docs/v3.

Chapter 9

1. Antonio, S. Amazon Sold This Warehouse to Hines in 2020. Amazon
Leases the Facility. https://www.hines.com/properties/6000-schertz-
parkway-schertz.
2. Statement on Businesswire. New MHI and Deloitte Report Finds
Supply Chain Disruption is Driving Big Investments in Technology.
https://www.businesswire.com/news/home/20220330005007/en/
New-MHI-and-Deloitte-Report-Finds-Supply-Chain-Disruption-is-
Driving-Big-Investments-in-Technology.
3. Supply Chain Game Changer. Machine Learning in Warehouse Management! https://supplychaingamechanger.com/machine-learning-
in-warehouse-management/.
4. Fedex. Innovation & Technology Policy Perspectives. https://www.
fedex.com/en-us/about/policy/technology-innovation.html.
5. Woyke, E. How UPS Uses AI to Deliver Holiday Gifts in the Worst
Storms. *MIT Technology Review*, 2018. https://www.technology

review.com/2018/11/21/139000/how-ups-uses-ai-to-outsmart-bad-weather/.

6. DHL. DHL Express Deploys Ai-Powered Sorting Robot. 2021, https://www.dhl.com/global-en/delivered/digitalization/ai-in-logistics.html.

7. Venkatesan, S. How Walmart is Using AI. To Make Smarter Substitutions in Online Grocery Orders. Walmart, 2021. https://corporate.walmart.com/newsroom/2021/06/24/headline-how-walmart-is-using-a-i-to-make-smarter-substitutions-in-online-grocery-orders.

8. Roberts, J.J. Walmart's Use of Sci-fi Tech to Spot Shoplifters Raises Privacy Questions. *Fortune*, 2015. https://fortune.com/2015/11/09/wal-mart-facial-recognition/.

9. Grill-Goodman, J. How Walmart Uses Computer Vision. RIS News, 2019. https://risnews.com/how-walmart-uses-computer-vision.

10. Sullivan, E. Amazon Fulfillment Center Locations: The Ultimate List. Tinuiti, 2022. https://tinuiti.com/blog/amazon/amazon-fulfillment-centers-map/.

11. The Story Behind Amazon's Next Generation Robot. Amazon, 2019. https://www.aboutamazon.com/news/innovation-at-amazon/the-story-behind-amazons-next-generation-robot.

12. AGV Robot. ALL the Amazon Warehouse Robots: From Fulfillment to Last-Mile Delivery. https://www.agvnetwork.com/robots-amazon.

13. Amazon Flex. https://flex.amazon.com/.

14. Feedvisor. How Amazon Leverages Artificial Intelligence to Optimize Delivery. October 2019. https://feedvisor.com/resources/amazon-shipping-fba/how-amazon-leverages-artificial-intelligence-to-optimize-delivery/.

15. *Ibid.*

16. Ocado. We Developed the Ocado Smart Platform (OSP) at the Intersection of Six Disruptive Technologies: AI, Robotics, Digital Twins, Cloud, Big Data, and IoT. https://www.ocadogroup.com/technology/powering-osp.

17. Ocado. Our Business Explained. https://www.ocadogroup.com/about-us/what-we-do/how-we-use-ai/.

18. Iddenden, G. Identity Crisis: Is Ocado a Tech Company or a Retailer? Charged, 2022. https://www.chargedretail.co.uk/2022/08/19/identity-crisis-is-ocado-a-tech-company-or-a-retailer/.
19. Kroger, S. N. Goes Live with Ocado 'Spoke' Site in Kentucky. August 2022. https://www.supermarketnews.com/online-retail/kroger-goes-live-ocado-spoke-site-kentucky.
20. Hopstack. Warehouse Transformation: The Future of Warehousing. August 2022. https://www.hopstack.io/blog/warehouse-transformation-the-future-of-warehousing.
21. Real Assets Adviser. Industrial's Next Act: The Future of Warehouse Lies in Automation and AI. September 2021.

Chapter 10

1. Girls of Steel Robotics. From Inspiration by R2D2 to Building Her Own Robots. January 2021. https://girlsofsteelrobotics.com/news/from-inspiration-by-r2d2-to-building-her-own-robots/.
2. European regulators blocked the deal and Amazon abandoned it in 2024.
3. Rus, Daniela. The machines from our future. *Daedalus*, 151(2), 100–113 (2022). https://direct.mit.edu/daed/article/151/2/100/110628/The-Machines-from-Our-Future.
4. Hardesty, L. Ingestible Origami Robot. MIT News, 2016. https://news.mit.edu/2016/ingestible-origami-robot-0512.
5. Gordon, R. Robot Hand is Soft and Strong. *MIT News*, 2019. https://news.mit.edu/2019/new-robot-hand-gripper-soft-and-strong-0315.
6. IRobot. The Setup Guide for a Roomba i Series and j Series Robot Vacuum. 2021. https://support.irobot.ca/s/article/32672.
7. Robust AI. Carter. https://www.robust.ai/blog/announcing-grace-and-carter.
8. Mordor Intelligence. Robotic Lawn Mower Market — Growth, Trends, COVID-19 Impact, And Forecasts (2022–2027). https://www.mordorintelligence.com/industry-reports/robotic-lawn-mower-market.

9. *Ibid.*
10. IRobot. Smart Home, Inside and Out, for 30 Years. https://www.irobot.com.au/roomba/innovations.
11. Google. Camera Alerts. https://support.google.com/googlenest/answer/9210305?hl=en.
12. Peshterliev, S. Active Learning: Algorithmically Selecting Training Data to Improve Alexa's Natural-language Understanding. Amazon Science, 2019. https://www.amazon.science/blog/active-learning-algorithmically-selecting-training-data-to-improve-alexas-natural-language-understanding.
13. Ackerman, D. System Brings Deep Learning to "Internet of Things" Devices. *MIT News,* 2020. https://news.mit.edu/2020/iot-deep-learning-1113.

Chapter 11

1. Stewart, M. Machine Learning Sensors: Truly Data-Centric AI. Towards Data Science, 2022. https://towardsdatascience.com/machine-learning-sensors-truly-data-centric-ai-8f6b9904633a.
2. Warden, P., and Situnayake, D. TinyML. O'Reilly. https://www.oreilly.com/library/view/tinyml/9781492052036/.
3. Hertz, J. What is TinyML? All About Circuits, 2022. https://www.allaboutcircuits.com/technical-articles/what-is-tinyml/.
4. Rus, Daniela, cited in Ford, M. *Architects of Intelligence: The Truth about AI from the People Building It.* UK, Packt Publishing, 2018.
5. Electrochemical Society. World of Sensors. https://www.electrochem.org/world-of-sensors.
6. Abraham, *et al. Smart Sensor Networks: Analytics, Sharing and Control.* Springer International Publishing, 2021.
7. *Ibid.,* p. 8.
8. Pete, W., *et al.* Machine Learning Sensors. https://arxiv.org/abs/2206.03266#:~:text=Abstract%3A%20Machine%20learning%20sensors%20represent,security%20concerns%20from%20data%20movement.

9. Bank My Cell. How Many Smartphones Are in the World? https://www.bankmycell.com/blog/how-many-phones-are-in-the-world#:~:text=How%20Many%20People%20Have%20Mobile%20Phones%20In%20The%20World%3F&text=In%202022%2C%20including%20both%20smart,the%20world%20cell%20phone%20owners.

10. Artificial Intelligence for Water Quality Monitoring. https://ecs.confex.com/ecs/aimes2018/meetingapp.cgi/Paper/112684.

11. *Machine Learning in Sensors and Imaging*. MDPI AG, 2022.

12. Ortiz, J. AI Powered Sensors Are the Future. Electrochemical Society, 2019. https://www.electrochem.org/ecsnews/AI-Powered-Sensors-Are-the-Future#:~:text=a%20sustainable%20world.-,AI%20powered%20sensors%20are%20the%20future.%E2%80%9D,bacteria%20based%20on%20visual%20appearance.

13. de Bruin, J. Future Sensor Technology: 21 Expected Trends. Sentech. https://www.sentech.nl/en/rd-engineer/21-sensor-technology-future-trends/.

Chapter 12

1. Benanav, A. *Automation and the Future of Work*. Verso Books, 2020. XI.

2. Emeritus. Pros and Cons of Automation in the Workplace. https://emeritus.org/blog/pros-and-cons-of-automation-in-the-workplace/.

3. Frey, C.B., and Osborne, M.A. The Future of Employment.Oxford University, 2013. https://www.oxfordmartin.ox.ac.uk/downloads/academic/The_Future_of_Employment.pdf.

4. Emeritus, *op. cit.*

5. UIPath. Brochure. https://www.uipath.com/resources/automation-brochures/ai-center-infographic-brochure.

6. McKinsey. Where Machines Could Replace Humans—And Where They Can't (Yet). 2016. https://www.mckinsey.com/business-functions/mckinsey-digital/our-insights/where-machines-could-replace-humans-and-where-they-cant-yet.

7. World Economic Forum. The Future of Jobs Report. 2018. https://www3.weforum.org/docs/WEF_Future_of_Jobs_2018.pdf.

8. Benanav, A. (2020), *op. cit.*
9. Andy, S. and Lee, K. *Raising the Floor: How a Universal Basic Income Can Renew Our Economy and Rebuild the American Dream.* 1st edn., Public Affairs, 2016.
10. Mogstad, M., and Kearney, M.S. Universal Basic Income (UBI) as a Policy Response to Current Challenges. 2019. https://www.economic strategygroup.org/publication/universal-basic-income-ubi-as-a-policy-response-to-current-challenges/.
11. Autor, D. Why are There Still So Many Jobs? The History and Future of Workplace Automation and Anxiety. https://ide.mit.edu/sites/default/files/publications/IDE_Research_Brief_v07.pdf.
12. Autor, D. Ted Talk. Will Automation Take Away All Our Jobs? https://www.youtube.com/watch?v=th3nnEpITz0.

Chapter 13

1. Lohr, S. IBM to Sell Watson Health Assets to Francisco Partners. *New York Times*, 2022. https://www.nytimes.com/2022/01/21/business/ibm-watson-health.html.
2. IBM's Rometty. Health Care Will Be Our Moon Shot. April 16, 2015. Charlie Rose. https://www.youtube.com/watch?v=46MYhalt7EU.
3. Memorial Sloan Kettering Cancer Center, IBM to Collaborate in Applying Watson Technology to Help Oncologists. https://www.mskcc.org/news-releases/mskcc-ibm-collaborate-applying-watson-technology-help-oncologists.
4. Lohr, S. What Ever Happened to IBM's Watson? *New York Times*, 2021. https://www.nytimes.com/2021/07/16/technology/what-happened-ibm-watson.html.
5. IBM Website. Press Release. January 2022. https://newsroom.ibm.com/2022-01-21-Francisco-Partners-to-Acquire-IBMs-Healthcare-Data-and-Analytics-Assets.
6. Kumar, Y., *et al.* Artificial intelligence in disease diagnosis: A systematic literature review, synthesizing framework and future research agenda. *Journal of Ambient Intelligence and Humanized Computing,*

1–28 (2022). doi:10.1007/s12652-021-03612-z. https://www.ncbi.
nlm.nih.gov/pmc/articles/PMC8754556/.

7. Suleyman, M. Announcing DeepMind Health Research Partnership
   with Moorfields Eye Hospital. DeepMind Blog, 2018. https://
   www.deepmind.com/blog/announcing-deepmind-health-research-
   partnership-with-moorfields-eye-hospital.

8. De Fauw, J., Ledsam, J. R., Romera-Paredes, B., *et al.* Clinically applicable deep learning for diagnosis and referral in retinal disease.
   Nature Medicine, 24, 1342–1350 (2018). https://doi.org/10.1038/
   s41591-018-0107-6.

9. DeepMind Blog. A Major Milestone for the Treatment of Eye
   Disease. https://www.deepmind.com/blog/a-major-milestone-for-the-
   treatment-of-eye-disease.

10. Zhang G., *et al.* Clinically relevant deep learning for detection
    and quantification of geographic atrophy from optical coherence
    tomography: A model development and external validation study.
    The Lancet. https://www.thelancet.com/journals/landig/article/
    PIIS2589-7500(21)00134-5/fulltext.

11. Digital Health. Moorfields Develops In-house Algorithm to Help
    Detect Geographic Atrophy. https://www.digitalhealth.net/2021/09/
    moorfields-develops-in-house-algorithm-to-help-detect-geographic-
    atrophy/.

12. Bresnick, J. Top 5 Use Cases for Artificial Intelligence in Medical
    Imaging. 2018. https://healthitanalytics.com/news/top-5-use-cases-
    for-artificial-intelligence-in-medical-imaging.

13. Freeman, T. Machine Learning Makes Its Mark on Medical Imag-
    ing and Therapy. Physics World, 2022. https://physicsworld.com/a/
    machine-learning-makes-its-mark-on-medical-imaging-and-
    therapy/.

14. Gould, M. K., *et al.* Machine learning for early lung cancer identifi-
    cation using routine clinical and laboratory data. *American Journal
    of Respiratory and Critical Care Medicine*, 204(4), 445–453 (2021).
    doi:10.1164/rccm.202007-2791OC. https://pubmed.ncbi.nlm.nih.
    gov/33823116/.

15. Heuvelmans, M., *et al.* Lung cancer prediction by deep learning to identify benign lung nodules. *Lung Cancer Journal.* https://www.sciencedirect.com/science/article/pii/S0169500221000453.

16. Mascia, K. How Data Science Is Ushering in a New Era of Modern Medicine. Johnson & Johnson, 2022. https://www.jnj.com/innovation/how-data-science-ushers-in-new-era-of-modern-medicine.

17. Silva, F., Pereira, T., Neves, I., Morgado, J., Freitas, C., Malafaia, M., Sousa, J., Fonseca, J., Negrão, E., Flor de Lima, B., Correia da Silva, M., Madureira, A. J., Ramos, I., Costa, J. L., Hespanhol, V., Cunha, A., Oliveira, H. P. Towards machine learning-aided lung cancer clinical routines: Approaches and open challenges. *Journal of Personalized Medicine*, 16; 12(3), 480 (2022). doi:10.3390/jpm12030480. PMID: 35330479; PMCID: PMC8950137.

18. Svoboda, E. Artificial Intelligence is Improving the Detection of Lung Cancer. *Nature*, 2020. https://www.nature.com/articles/d41586-020-03157-9#:~:text=Going%20deeper,of%206%20veteran%20radiologists1.

19. Krittanawong, C., Virk, H. U. H., Bangalore, S., *et al.* Machine learning prediction in cardiovascular diseases: A meta-analysis. *Scientific Reports*, 10, 16057 (2020). https://doi.org/10.1038/s41598-020-72685-1; https://www.nature.com/articles/s41598-020-72685-1.

20. *Ibid.*

21. Varoquaux, G., Cheplygina, V. Machine learning for medical imaging: Methodological failures and recommendations for the future. *npj Digital Medicine*, 5, 48 (2022). https://doi.org/10.1038/s41746-022-00592-y.

22. Stanford University HAI. Artificial Intelligence Index Report 2022. https://aiindex.stanford.edu/wp-content/uploads/2022/03/2022-AI-Index-Report_Master.pdf.

23. Kurzweil. By 2030, Nanobots Will Flow Throughout Our Bodies. Futurism. https://futurism.com/kurzweil-by-2030-nanobots-will-flow-throughout-our-bodies.

Chapter 14

1. IDC. The Digitization of the World. Deloitte on Market Size. https://www2.deloitte.com/cy/en/pages/technology/articles/data-grown-big-value.html.
2. O'Neil, C. *Weapons of Math Destruction: How Big Data Increases Inequality and Threatens Democracy.* UK, Crown, 2016.
3. Boddington, P. *Toward a Code of Ethics for AI.* Springer International Publishing, 2017.
4. Stahl, B. C. Ethical issues of AI. *Artificial Intelligence for a Better Future*, 18, 35–53 (2021).
5. Pellerin, C. Project Maven. Department of Defense, 2017. https://www.defense.gov/News/News-Stories/Article/Article/1254719/project-maven-to-deploy-computer-algorithms-to-war-zone-by-years-end/.
6. Google Employee Letter to CEO. https://static01.nyt.com/files/2018/technology/googleletter.pdf.
7. Smith, B. Technology and the Military. Microsoft, 2018. https://blogs.microsoft.com/on-the-issues/2018/10/26/technology-and-the-us-military/.
8. Kelion, L. Huawei Patent Mentions Use of Uighur-spotting Tech. *BBC*, 2021. https://www.bbc.com/news/technology-55634388.
9. Wolford, B. What is GDPR, the EU's New Data Protection Law? European Union. https://gdpr.eu/what-is-gdpr/.
10. Schwartz, O. In 2016, Microsoft's Racist Chatbot Revealed the Dangers of Online Conversation. IEEE Spectrum, 2019. https://spectrum.ieee.org/in-2016-microsofts-racist-chatbot-revealed-the-dangers-of-online-conversation.
11. Google Apologises for Photos app's Racist Blunder. *BBC News*, 2015. https://www.bbc.com/news/technology-33347866.
12. Coeckelbergh, M. *AI Ethics.* Cambridge, MA, USA, The MIT Press, 2020.

Chapter 15

1. Athey, S. NBER. *The Economics of Artificial Intelligence: An Agenda.* University of Chicago Press. May 2019. https://www.nber.org/system/files/chapters/c14009/c14009.pdf.
2. Stanford. How will AI Change Jobs? https://hai.stanford.edu/.
3. DeepMind Blog, 2021. https://www.deepmind.com/blog/nowcasting-the-next-hour-of-rain#:~:text=We%20focus%20on%20nowcasting%20rain,problem%20of%20generating%20radar%20movies.
4. Atlanta Fed. GDP Now. https://www.atlantafed.org/cqer/research/gdpnow?panel=2.
5. Bolívar, F., *et al.* A Big Data Approach to Understand Central Banks. BBVA, November 2018. https://www.bbvaresearch.com/en/publicaciones/a-big-data-approach-to-understand-central-banks/.
6. Doerr, S., *et al.* Big Data and Machine Learning in Central Banking. Bank for International Settlements, 2021. https://www.bis.org/publ/work930.pdf.
7. Bang, J. T., *et al. Machine-learning Techniques in Economics: New Tools for Predicting Economic Growth.* Germany, Springer International Publishing, 2017.
8. Athey, (2019), *op. cit.*
9. Brynjolfsson, E. The Turing Trap: The Promise & Peril of Human-Like Artificial Intelligence. Stanford Digital Economy Lab, 2022. https://digitaleconomy.stanford.edu/news/the-turing-trap-the-promise-peril-of-human-like-artificial-intelligence/.
10. Haponik, A. Machine Learning in Economics. Addepto, 2019. https://addepto.com/machine-learning-in-economics-how-is-it-used/.

Chapter 16

1. Zuckerman, G. *The Man Who Solved the Market.* Penguin Random House, 2019.

2. Machine Learning in Hedge Fund Investing. JPMorgan Asset Management, 2019. https://am.jpmorgan.com/hu/en/asset-management/institutional/insights/portfolio-insights/machine-learning-in-hedge-fund-investing/.
3. Ted Talk interview with Jim Simons and Chris Anderson. 2016. https://www.youtube.com/watch?v=U5kIdtMJGc8.
4. Zuckerman (2019), *op. cit.*
5. Nagel, S. *Machine Learning in Asset Pricing*. Princeton, Princeton University Press, 2021.
6. *Ibid.*, p. 6.
7. Man Group. The Rise of Machine Learning. November 2016. https://www.man.com/maninstitute/the-rise-of-machine-learning.
8. Goldman Sachs Asset Management. The Role of Big Data in Investing. https://www.gsam.com/content/gsam/global/en/market-insights/gsam-insights/gsam-perspectives/2016/big-data/gsam-roundtable.html.

Chapter 17

1. Anti-Money Laundering and Countering the Financing of Terrorism. US State Department Policy. https://www.state.gov/anti-money-laundering-and-countering-the-financing-of-terrorism.
2. Munoz, P. How Machine Learning Can Help Fight Money Laundering. Capital One, 2021. https://www.capitalone.com/tech/machine-learning/how-machine-learning-can-help-fight-money-laundering/.
3. Biswas, S., *et al.* AI-bank of the Future: Can Banks Meet the AI Challenge? McKinsey, 2020. https://www.mckinsey.com/industries/financial-services/our-insights/ai-bank-of-the-future-can-banks-meet-the-ai-challenge.
4. Bank of England and FCA. Machine Learning in UK Financial Services. October 2019. https://www.bankofengland.co.uk/report/2019/machine-learning-in-uk-financial-services.

5. Opus. Rule-Based vs. Machine Learning: Effective Fraud Prevention Models. October 2021. https://www.opusconsulting.com/rule-based-vs-machine-learning-effective-fraud-prevention-models/#:~:text=Rules%2Dbased%20requires%20multiple%20steps,buried%20within%20subtle%20pattern%20changes.

6. Institute of International Finance. Machine Learning in Anti-Money Laundering — Summary Report, p. 4. October 2018. https://www.iif.com/portals/0/Files/private/32370132_iif_machine_learning_in_aml_-_public_summary_report.pdf.

7. *Ibid.*

8. Bank of England Quarterly Bulletin. Q4 2020. https://www.bankofengland.co.uk/quarterly-bulletin/2020/2020-q4/the-impact-of-covid-on-machine-learning-and-data-science-in-uk-banking.

9. Bank of England. Agenda for Research. https://www.bankofengland.co.uk/research/bank-of-england-agenda-for-research.

10. Bank of America. Erica. https://promotions.bankofamerica.com/digitalbanking/mobilebanking/erica.

11. Bank of America Press Release.

12. Wells Fargo. The Most Helpful 'Banking Assistant' on Facebook. July 2017. https://stories.wf.com/helpful-banking-assistanton-facebook/#:~:text=Wells%20Fargo%20has%20received%20industrywide,Fargo's%20helpful%20virtual%20banking%20assistant.

13. JPMorgan. Press Release. June 2018. https://www.jpmorgan.com/news/jpmorgan-to-pilot-an-aipowered-virtual-assistant.

14. Chatbot Guide. JPMorganChase. https://www.chatbotguide.org/jpmorgan-bot.

15. Visa. Visa Introduces AI-Powered Innovations for Smarter Payments. February 2021. https://usa.visa.com/about-visa/newsroom/press-releases.releaseId.17701.html.

16. Hedgeweek. New York Institute of Finance and Google Cloud Launch a Machine Learning for Trading Specialisation on Coursera. January 2020. https://www.hedgeweek.com/2020/01/27/282340/new-york-institute-finance-and-google-cloud-launch-machine-learning-trading.

17. CFA Institute. Programs. https://www.cfainstitute.org/en/programs/cfa/charterholder-careers/roles/data-scientist.

Chapter 18

1. Standage, T. *A Brief History of Motion: From the Wheel to the Car to What Comes Next.* Bloomsbury Publishing, 2021.
2. NHTSA. Levels of Automation. May 2022. https://www.nhtsa.gov/sites/nhtsa.gov/files/2022-05/Level-of-Automation-052522-tag.pdf.
3. NHTSA. Automated Vehicle Safety. https://www.nhtsa.gov/technology-innovation/automated-vehicles-safety.
4. Traffic Safety Store. https://www.trafficsafetystore.com/blog/automotive-autonomy/.
5. Tesla. Artificial Intelligence & Autopilot. https://www.tesla.com/AI.
6. Futurism. Report: Tesla Crashes Dropped by 40% After Autopilot was Installed. January 2017. https://futurism.com/report-tesla-crashes-dropped-by-40-after-autopilot-was-installed.
7. NHTSA Report. 2017. https://static.nhtsa.gov/odi/inv/2016/INCLA-PE16007-7876.PDF.
8. Marr, B. *Artificial Intelligence in Practice.* Chichester, Wiley, 2019, p. 266.
9. *Ibid.*, p. 289.
10. Standage, (2021), *op. cit.*
11. *Ibid.*
12. Barra, M. Post on LinkedIn. October 2017. https://www.linkedin.com/pulse/zero-crashes-emissions-congestion-mary-barra/.
13. US Department of Transportation. https://www.transportation.gov/av/avcp/5; https://www.transportation.gov/AV.
14. Coeckelbergh, M. *AI Ethics.* Cambridge, MA, USA, The MIT Press, 2020.
15. Allianz. Autonomous Trains. https://commercial.allianz.com/news-and-insights/expert-risk-articles/autonomous-trains.html.
16. Waymo. Waymo Driver explanation. https://waymo.com/waymo-driver/.

17. Volvo. Press Release. September 2021.
18. Maritime Executive. Yara Birkeland Begins Further Testing for Autonomous Operations. April 2022. https://maritime-executive. com/article/yara-birkeland-christened-and-begins-testing-for-autonmous-operations.
19. Kamps, H.J. MIT's CSAIL self-driving water taxis launched in the Amsterdam canals. TechCrunch, 2021. https://techcrunch.com/ 2021/10/29/mit-csail/#:~:text=This%20week%2C%20scientists%20 from%20MIT's,voyages%20in%20the%20canals%20today.
20. Gordon, R. Autonomous boats could be your next ride. MIT, 2020. https://news.mit.edu/2020/autonomous-boats-could-be-your-next-ride-1026; https://roboat.org/.

Chapter 19

1. Marcus, G., Davis, E. *Rebooting AI: Building Artificial Intelligence We Can Trust.* 1st edn., Pantheon Books, 2019.
2. Saha, B. Machine Learning is Moving beyond the Hype. Infoworld, 2021. https://www.infoworld.com/article/3630517/machine-learning-is-moving-beyond-the-hype.html.
3. https://aiindex.stanford.edu/wp-content/uploads/2022/03/2022-AI-Index-Report_Master.pdf.

Chapter 20

1. Botto's Website. https://www.botto.com/press.
2. Audry, S. *Art in the Age of Machine Learning.* MIT Press, 2021.
3. MOMA. Identifying Art through Machine Learning. https://www. moma.org/calendar/exhibitions/history/identifying-art.
4. Botto. Genesis Period. https://www.botto.com/gallery/genesis-period.
5. Audry, (2021), *op. cit.*, p. 18.
6. Aguera y Arcas, B. Artists + Machine Intelligence, February 2016. https://medium.com/artists-and-machine-intelligence/what-is-ami-96cd9ff49dde.

7. OpenAI. Dall-E 2. https://openai.com/dall-e-2/.
8. Cheng, I. Minimum Viable Sentience. 2020. http://iancheng.com/minimumviablesentience.
9. Beeple Website. https://www.beeple-crap.com/about.
10. Beeple Tweet. https://twitter.com/beeple/status/1559181938581200897.
11. Rutgers. The Art and Artificial Intelligence Laboratory at Rutgers: Advancing AI Technology in the Digital Humanities. https://sites.google.com/site/digihumanlab/home.
12. Elgammal, A. Generating "art" by Learning about Styles and Deviating from Style Norms. https://medium.com/@ahmed_elgammal/generating-art-by-learning-about-styles-and-deviating-from-style-norms-8037a13ae027.
13. Aican. Art of the Future, Now. https://www.aican.io.
14. Elgammal, A. Meet Aican. *Interalia Magazine*, 2019. https://www.interaliamag.org/articles/ahmed-elgammal/.
15. Naskar, A. Humanizing AI (The Sonnet) | Abhijit Naskar | Either Reformist or Terrorist. https://naskarism.wordpress.com/2022/04/09/humanizing-ai-the-sonnet-abhijit-naskar-either-reformist-or-terrorist/.

Chapter 21

1. Lee, K.-F., Qiufan, C. *AI 2041: Ten Visions for Our Future*. UK, Ebury Publishing, 2021.
2. Grace, P. How AI and Machine Learning are Transforming the Education Sector. *AACE Review*, 2021. https://www.aace.org/review/ai-and-machine-learning/.
3. Coursera. How Coursera Works. https://about.coursera.org/how-coursera-works/.
4. Ng, A. Tweet, April 2022. https://twitter.com/AndrewYNg/status/1516090521281724417?ref_src=twsrc%5Etfw%7Ctwcamp%5Etweetembed%7Ctwterm%5E1516090521281724417%7Ctwgr%5E%7Ctwcon%5Es1_&ref_url=https%3A%2F%2Fanalyticsindiamag.com%2Fandrew-ng-to-teach-updated-version-of-his-popular-ml-specialization-course%2F.

5. Lynch, M. 6 Ways Machine Learning Will Revolutionize the Education Sector. The Tech Advocate, 2019. https://www.thetechedvocate.org/6-ways-machine-learning-will-revolutionize-the-education-sector/.

6. Amazon. Machine Learning in Education. https://aws.amazon.com/education/ml-in-education/.

7. AACE, *op. cit.*

8. Khan, I., Ahmad, A. R., Jabeur, N., *et al.* An artificial intelligence approach to monitor student performance and devise preventive measures. *Smart Learning Environments*, 8, 17 (2021). https://doi.org/10.1186/s40561-021-00161-y.

9. Broussard, M. *Artificial Unintelligence: How Computers Misunderstand the World.* The MIT Press, 2018, p. 68.

10. O'Neil, C. *Weapons of Math Destruction: How Big Data Increases Inequality and Threatens Democracy.* UK, Crown, 2016.

11. Jimenez, L., and Boser, U. Future of Testing in Education: Artificial Intelligence. Center for American Progress, 2021. https://www.americanprogress.org/article/future-testing-education-artificial-intelligence/.

12. Ofgang, E. How Machine Learning is having an Impact on Education. *Tech & Learning*, 2021. https://www.techlearning.com/news/how-machine-learning-is-having-an-impact-on-education.

13. Jimenez, L., and Boser, U. (2021), *op. cit.*

14. AACE, *op. cit.*

15. Habib (ed.). *Revolutionizing Education in the Age of AI and Machine Learning.* USA, IGI Global, 2019.

Chapter 22

1. Dunn, J. Introducing FBLearner Flow: Facebook's AI backbone. Facebook, 2016. https://engineering.fb.com/2016/05/09/core-data/introducing-fblearner-flow-facebook-s-ai-backbone/.

2. *Artificial Intelligence in Social Media.* IntroBooks.

3. Bullock, J., & Korinek, A. @elonmusk and @twitter: The Problem with Social Media is Misaligned Recommendation Systems, not

free speech. Brookings, 2022. https://www.brookings.edu/research/
elonmusk-and-twitter-the-problem-with-social-media-is-misaligned-
recommendation-systems-not-free-speech/.

4. Mosseri, A. Our Commitment to Lead the Fight against
   Online Bullying. Instagram, 2019. https://about.instagram.com/
   blog/announcements/instagrams-commitment-to-lead-fight-
   against-online-bullying.

5. Instagram: Bully filter. May 2018. https://about.instagram.com/blog/
   announcements/bully-filter-and-kindness-prom-to-protect-our-
   community.

6. Systrom, K. Self-expression. Instagram, 2017. https://about.instagram.
   com/blog/announcements/keeping-instagram-a-safe-place-for-self-
   expression.

7. Card, C. How Facebook AI Helps Suicide Prevention. Facebook
   Website, 2018. https://about.fb.com/news/2018/09/inside-feed-suicide-
   prevention-and-ai/.

8. Marr, 2019, p. 159.

9. Qi He. Building the LinkedIn Knowledge Graph. LinkedIn, 2016.
   https://engineering.linkedin.com/blog/2016/10/building-the-
   linkedin-knowledge-graph#:~:text=LinkedIn's%20knowledge%
   20graph%20is%20a,geographical%20locations%2C%20schools%
   2C%20etc.

10. https://www.microsoft.com/en-us/investor/earnings/fy-2022-q3/
    press-release-webcast.

11. Dev.To. How does TikTok use machine learning? https://dev.to/
    mage_ai/how-does-tiktok-use-machine-learning-5b7i.

12. TikTok Open Careers Page. https://careers.tiktok.com/.

13. Klubnikin, A. 3 reasons to create a dating app with AI features. Itrex,
    2021. https://itrexgroup.com/blog/ai-for-dating-apps/.

14. Ng, E. OkCupid Tech Blog, 2020. https://tech.okcupid.com/vespa-
    vs-elasticsearch-for-matching-millions-of-people-6e3af18eb4dc.

15. Belloni, M. Multilingual message content moderation at scale.
    Bumble, 2021. https://medium.com/bumble-tech/multilingual-
    message-content-moderation-at-scale-ddd0da1e23ed.

16. Scanlon, K. Dating Data: An Overview of the Algorithm. The Startup, 2020. https://medium.com/swlh/dating-data-an-overview-of-the-algorithm-afb9f0c08e2c#:~:text=Hinge%20uses%20the%20Gale%2DShapley,to%20people%20with%20similar%20preferences.

17. Tao, L., and Deng, I. China's Tinder embraces AI as it eyes growth from the country's singles. *South China Morning Post*, 2018. https://www.scmp.com/tech/china-tech/article/2154856/chinas-tinder-embraces-ai-it-eyes-growth-countrys-singles.

18. Bengio, Yoshua, cited in Ford, Martin. *Architects of Intelligence: The Truth about AI from the People Building It.* UK, Packt Publishing, 2018.

Chapter 23

1. Amazon. How Does Alexa Work? https://www.amazon.com/how-does-alexa-work/b?ie=UTF8&node=21166405011.

2. Marr, B. Machine Learning in Practice: How Does Amazon's Alexa Really Work? https://bernardmarr.com/machine-learning-in-practice-how-does-amazons-alexa-really-work/#:~:text=So%2C%20when%20you%20ask%20Alexa,output%20back%20to%20your%20device.

3. Amazon Lex. https://aws.amazon.com/lex/.

4. Google. How Google Assistant Works. https://developers.google.com/assistant/howassistantworks.

5. Sullivan, D. Google Assistant Guide. Search Engine Land, 2017. https://searchengineland.com/google-assistant-guide-270312#:~:text=Google%20Assistant%20is%20activated%20by,Google%E2%80%9D%20followed%20by%20your%20query.

6. Google. Dialogflow. https://cloud.google.com/dialogflow/docs/release-notes.

7. Apple. Siri. https://www.apple.com/siri/.

8. Apple. Machine Learning Research. https://machinelearning.apple.com/research/hey-siri#:~:text=The%20%22Hey%20Siri%22%20detector%20uses,uttered%20was%20%22Hey%20Siri%22.

9. Apple Support. https://support.apple.com/guide/iphone/ask-siri-iph83aad8922/ios#:~:text=When%20you%20activate%20Siri%20with,an%20alarm%20for%208%20a.m.%E2%80%9D.

10. Warren, T. Microsoft no Longer Sees Cortana as an Alexa or Google Assistant Competitor. The Verge, 2019. https://www.theverge.com/2019/1/18/18187992/microsoft-cortana-satya-nadella-alexa-google-assistant-competitor.

11. Gartner. Brace Yourself for an Explosion of Virtual Assistants. August 2020. https://blogs.gartner.com/anthony_bradley/2020/08/10/brace-yourself-for-an-explosion-of-virtual-assistants/#:~:text=Gartner%20predicts%20that%20by%202025,up%20from%202%25%20in%202019.

12. Lellouche Tordjman, K. Ted Talk on Virtual Assistants. https://www.ted.com/talks/karen_lellouche_tordjman_siri_alexa_google_what_comes_next?language=en.

13. Mulley-Goodbarne, E. The Future of Virtual Assistants Might lie in the Metaverse. IT Pro, 2022. https://www.itpro.com/technology/voice-assistant/367610/future-of-virtual-assistants-lies-in-the-metaverse.

14. Amazon's big dreams for Alexa fall short. *Financial Times*. March 2023. https://www.ft.com/content/bab905bd-a2fa-4022-b63d-a385c2a0fb86.

Chapter 24

1. Korolov, M. 9 Ways Hackers will Use Machine Learning to Launch Attacks. CSO, 2022. https://www.csoonline.com/article/3250144/6-ways-hackers-will-use-machine-learning-to-launch-attacks.html.

2. Brewster, T. Fraudsters Cloned Company Director's Voice In $35 Million Bank Heist, Police Find. *Forbes*, 2021. https://www.forbes.com/sites/thomasbrewster/2021/10/14/huge-bank-fraud-uses-deep-fake-voice-tech-to-steal-millions/?sh=88212ac75591.

3. Citrix. What is Access Control? https://www.citrix.com/solutions/secure-access/what-is-access-control.html.

4. Cisco. How Machine Learning Helps Security. https://www.cisco.com/c/en/us/products/security/machine-learning-security.html#~how-ml-helps-security.

5. Chio, C., Freedman, D. *Machine Learning & Security*. Sebastopol, O'Reilly Media, 2018.

6. Nemire, B. Technical Blog. Nvdia, 2016. https://developer.nvidia.com/blog/share-your-science-visualizing-200m-cybersecurity-alerts-daily-with-gpus/.

7. Castelli, J. Why Machine Learning Is a Critical Defense Against Malware. Crowdstrike, 2019. https://www.crowdstrike.com/blog/defending-against-malware-with-machine-learning/.

8. Jarvis, T. Autonomous Response Stops a Runaway Trickbot Intrusion Darktrace, 2022. https://darktrace.com/blog/autonomous-response-stops-a-runaway-trickbot-intrusion.

9. FBI/CISA Notice. March 2021. https://www.cisa.gov/uscert/ncas/alerts/aa21-076a.

10. Gottsegen, G. Machine Learning Cybersecurity: How It Works and Companies to Know. BuiltIn, 2019. https://builtin.com/artificial-intelligence/machine-learning-cybersecurity.

11. *Ibid.*

12. DOJ Release. September 2018. https://www.justice.gov/opa/pr/north-korean-regime-backed-programmer-charged-conspiracy-conduct-multiple-cyber-attacks-and.

13. Georgetown CSET. Machine Learning and Cybersecurity Hype and Reality. June 2021. https://cset.georgetown.edu/publication/machine-learning-and-cybersecurity/.

14. Russell, S. *The Reith Lectures 2021*. AI In Warfare. https://www.bbc.co.uk/programmes/m00127t9.

15. Chio and Freedman. (2018), *op. cit.*

Chapter 25

1. UNESCO Report on AI. https://ircai.org/wp-content/uploads/2022/03/IRCAI-2021-Annual-Artificial_Intelligence-SDGs-TOP-100-Report.pdf.

2. Oxford University. Spotting Elephants from Space: A Satellite Revolution. December 2020. https://www.ox.ac.uk/news/2020-12-18-spotting-elephants-space-satellite-revolution#:~:text=Remotely%20sensing%20elephants%20using%20satellite,the%20risk%20of%20double%20counting.

3. Wahed, A. Among the Elephants Blog May 2020. https://www.save theelephants.org/blog/?detail=artificial-intelligence-for-elephants& amp=1. Save the Elephants. Tracking — Real Time Monitoring. https://www.savetheelephants.org/project/tracking-real-time-monitoring-app/.

4. WildMe. Conservation Meets Machine Learning. https://www.wildme.org/#/.

5. Schlossberg, T. AI is Helping Scientists Understand an Ocean's Worth of Data. *New York Times*, 2020. https://www.nytimes.com/2020/04/08/science/ai-ocean-whales-study.html.

6. Fei-Fei Li and Jia Li. Cloud AutoML: Making AI Accessible to Every Business. https://cloud.google.com/blog/topics/inside-google-cloud/cloud-automl-making-ai-accessible-every-business.

7. Cornell Lab. Elephant Listening Project. https://elephantlistening-project.org/landscapes/. Conservation Metrics. https://conservation metrics.com/.

8. Tuia, D., *et al.* Perspectives in Machine Learning for Wildlife Conservation. *Nature*, 2022. https://www.nature.com/articles/s41467-022-27980-y#Sec17.

9. WWF. Explore the Power of Machine Learning and AI. https://techhub.wwf.ca/technology/ai-machine-learning-applications/.

10. UN Environment Programme. UNEP and Google Partner to Hunt for Plastic Pollution with Machine Learning. April 2021. https://www.unep.org/news-and-stories/press-release/unep-and-google-partner-hunt-plastic-pollution-machine-learning.

11. Global Plastic Watch. https://globalplasticwatch.org/.

12. Industry Europe. Using Satellites & AI to Measure Plastic Pollution. May 2022. https://industryeurope.com/sectors/chemicals-biochemicals/using-satellites-ai-to-measure-plastic-pollution/.

13. Learn, J.R. AI Image Recognition. The Wildlife Society, 2022. https://wildlife.org/tws2021-ai-image-recognition-could-be-the-future-of-wildlife-management/.
14. WiseEye. Smart Feeders. https://wiseeyetech.com/.
15. Petersen, S., *et al.* Ecological Research. DeepMind, 2019. https://www.deepmind.com/blog/using-machine-learning-to-accelerate-ecological-research.
16. Wearn, O.R., Freeman, R., and Jacoby, D.M.P. Responsible AI for Conservation. *Nature*, 2019. https://www.nature.com/articles/s42256-019-0022-7.
17. Tuia, D., *et al.* Perspectives in Machine Learning for Wildlife Conservation. *Nature*, 2022. https://www.nature.com/articles/s41467-022-27980-y.

Chapter 26

1. Lee, A. What Are Large Language Models Used For? Nvidia Blog, 2023. https://blogs.nvidia.com/blog/2023/01/26/what-are-large-language-models-used-for.
2. Chow, A.R. How ChatGPT Managed to Grow Faster Than TikTok or Instagram. *Time*, 2023. https://time.com/6253615/chatgpt-fastest-growing.
3. Griffith, E, and Metz, C. 'Let 1,000 Flowers Bloom': AI. Funding Frenzy Escalates. *New York Times*, 2023. https://www.nytimes.com/2023/03/14/technology/ai-funding-boom.html.
4. Keel, M. AI Chat Bots Spout Misinformation and Hate Speech. Rolling Stone, 2023. https://www.rollingstone.com/culture/culture-features/ai-chat-bots-misinformation-hate-speech-1234677574/.
5. Open AI Website.
6. PWC Press Release.
7. OpenAI Website.
8. *Ibid.*
9. Duolingo Blog.
10. Gewirtz, D. How to use ChatGPT to write code. ZD Net, 2023. https://www.zdnet.com/article/how-to-use-chatgpt-to-write-code/.

11. Microsoft Website. https://microsoft.github.io/prompt-engineering/.
12. Bushwick, S. What the New GPT-4 AI Can Do. Scientific American, 2023. https://www.scientificamerican.com/article/what-the-new-gpt-4-ai-can-do/.
13. Popli, N. GPT-4 Has Been Out for 1 Day. These New Projects Show Just How Much More Powerful It is. *Time*, 2023. https://time.com/6263475/gpt4-ai-projects/.
14. Browne, R. Italy Became the First Western Country to Ban ChatGPT. Here's what other countries are doing. *CNBC Report*, 2023. https://www.cnbc.com/2023/04/04/italy-has-banned-chatgpt-heres-what-other-countries-are-doing.html.
15. Robertson, A. Italy later lifted the ban. The Verge, 2023. https://www.theverge.com/2023/4/28/23702883/chatgpt-italy-ban-lifted-gpdp-data-protection-age-verification.
16. Kishida Says Japan Aims to Lead Int'l Rule-making Efforts for AI Use. *Kyodo News*, 2023. https://english.kyodonews.net/news/2023/05/9a5944b4570d-kishida-says-japan-aims-to-lead-intl-rule-making-efforts-for-ai-use.html.

Conclusion

1. Hassabis, D. AlphaFold Reveals the Structure of the Protein Universe. DeepMind Blog, 2022. https://www.deepmind.com/blog/alphafold-reveals-the-structure-of-the-protein-universe.
2. Callaway, E. 'It will Change Everything': DeepMind's AI Makes Gigantic Leap in Solving Protein Structures. *Nature*, 2020. https://www.nature.com/articles/d41586-020-03348-4.
3. Metz, C. 'The Godfather of AI.' Leaves Google and Warns of Danger Ahead. *New York Times*, 2023. https://www.nytimes.com/2023/05/01/technology/ai-google-chatbot-engineer-quits-hinton.html.

Index

bias, 95, 99
bias and inequality, 144
bidirectional encoder
  representations from
  transformers (BERTs), 22, 52
big data, 95, 147, 169
big-data specialists, 107
Bing, 21
biometrics, 33
black swan events, 111
blind people, 37
Bloomberg LP, 51
Bostrom, Nick, xiv, 97
bubble, 129
business analytics, 129
business problems, 130
ByteDance, 150

C
Cambridge Analytica, 99
cameras, 36
can machines create art?, 133
Captcha, 29
Casinos, 38
CFA Institute, 119
chatbots, 148, 154, 171
ChatGPT, 49, 156, 172
China, 37
clear language, 130
clerical work, 81
clustering algorithms, 13
computer models, 63
computer programmers, 116
Computer Science and Artificial
  Intelligence Laboratory (CSAIL),
  126
computer scientists, 130
computer vision, 26, 56, 62, 150

conservation, 170
convolutional neural networks
  (CNNs), 6, 28, 30, 91, 166
Cortana, 155
Coursera, 143
crime, 118
Crowdstrike, 161
customers, 152
customer service and chatbots, 118
cybercriminals, 160
cybersecurity, 162
cycles of boom and bust, 129

D
Dall-E, 136
data analysis, 85, 127
data architecture, 131
data engineering, 85
data engineers, 130
data management, 85
data mining, 150
data modeling, 114
data science, 85
data scientists, 72, 116, 129
dating apps, 150
Dean, Jeff, 57
deep fake, 160
deep learning, 2, 20, 23, 36, 127
DeepMind, 90, 97, 177
deep neural networks, 15
definition: Machine learning, 1
Deloitte, 84
detect anomalies, 109
diagnostic medicine, 88
digital strategies, 85
distributed reinforcement learning,
  41
driverless vehicles, 124

Printed in the United States
by Baker & Taylor Publisher Services